STOP THE HASSLE

SIMPLIFY, SATISFY, AND SUCCEED

JIM BRAMLETT

Special Edition: includes the Hassle Score Survey
& Transformative Pillars for Business Growth

Stop the Hassle: Simplify, Satisfy, and Succeed

Printed in the United States of America

Paperback ISBN: 978-1-965319-00-0
eBook ISBN: 978-1-965319-01-7

Purpose Publishing

Purpose Publishing LLC.
13194 US Highway 301 South
Suite 417
Riverview, Florida 33578
www.PurposePublishing.com

Contents

Book Dedication

I am dedicating this book to those business leaders who risk everything and sacrifice so much to grow companies. The growth of companies is essential to give others employment and purpose. It's extremely difficult leading a company, a department, and people. I also dedicate this book to my family, who sacrificed so much to allow me to chase my dreams a couple of times. If anyone can take nuggets from this book, apply them, and grow, then my purpose will have been served.

Introduction

Introduction

Hello, I'm Jim Bramlett, and over the past four decades, I've learned a thing or two about running successful businesses. One key lesson I've learned is the importance of removing hassles for customers. Whether it's making the buying process more convenient, offering competitive prices, enhancing the product experience, or building trust, simplifying the customer experience can lead to increased loyalty and profitability.

In this introduction, I'll share some of my insights and experiences in these four core areas and introduce you to the **Hassle Score**, a practical tool for understanding and addressing customer hassles. Before discussing this, let's cover crucial points that supported my research in creating this score.

Convenience

Imagine this: you're rushing to get to work, and you realize you're out of coffee. Instead of making a detour to the store, you remember that there's a coffee subscription service that delivers your favorite blend right to your doorstep. This level of convenience is what customers value most. In my businesses, I've always looked for ways to make the buying process as easy and seamless as possible. Whether it's offering online ordering, fast shipping, or hassle-free returns, convenience is key to customer satisfaction.

Price

Another key factor in customer satisfaction is pricing. Customers want to get a good deal, but they also want to know that they're getting a quality product. Besides competitive pricing, buyers seek transparency, or knowing exactly the total cost of a product or service before they buy. Buyers don't like to be nickel-and-dimed with fine print, obscure charges and ancillary fees.

One of my successful ventures was an internet-based third-party logistics firm. We were able to bundle the shipping volume of many so that everyone enjoyed lower pricing than they could negotiate on their own. We also were able to change processes for certain shippers to lower overall cost. In each case, we ensured each customer knew their exact shipping charges before they shipped their product and provided a much simpler and comprehensive pricing format than what the industry had become accustomed to.

Product Experience

The product experience is another area where removing hassles can make a big difference. Customers want products that are easy to use, reliable, and meet their needs. One way I've enhanced the product experience in my businesses is by investing in product design and development.

For example, when I launched a SaaS-based platform to simplify data integrations between shipper's last mile carriers, we enabled shipping customers to receive real-time status updates on their shipments. By focusing on the end-user experience, we enabled sellers of merchandise, especially those shipping to residences to provide a much better delivery experience and eliminated many phone calls and emails inquiring about shipment status.

Trust

Trust is perhaps the most important factor in building customer loyalty. Customers need to trust that you will deliver on your promises and stand behind your products. One way I've built trust with my customers is by being transparent and honest in my dealings. Whether it's pricing, policies, or product information, I've always been upfront with my customers. This has helped me build a reputation for integrity.

Earned Wisdom on Reducing Hassles

Throughout my four-decade business journey, I've started five new services and worked for and founded four different companies. Initially, I didn't fully grasp the critical components of customer experience. However, through time and study, I came to understand what buyers truly seek and how to fulfill those desires. I learned that competition is fierce, and to succeed, a seller must embrace a culture of constant improvement to meet customer needs.

Understanding Customer Desires

When I first started in business, I thought success was all about offering a great product at a competitive price. However, I quickly realized that customers are looking for more than just a good deal. They want a seamless, hassle-free experience from start to finish. This includes everything from the buying process to customer service and beyond. By focusing on these aspects, I differentiated my businesses and built a loyal customer base.

Creating a Culture of Improvement

Another critical lesson I've learned is the importance of constantly improving your offerings. Competition is relentless, and if you're not evolving, you're falling behind. That's why I've always strived to create a culture where everyone in the organization is aligned to improve upon the ingredients that buyers are seeking. This means not only listening to customer feedback but also proactively seeking ways to enhance the customer experience.

Embracing Change

In today's fast-paced world, the only constant is change. As a business owner, it's crucial to embrace change and adapt to new technologies and trends. This might mean updating your website, offering new services, or revamping your marketing strategy. By staying ahead of the curve, you can ensure that your business remains relevant and competitive in the marketplace.

The Role of Customer Feedback

Customer feedback is invaluable for improving your business. I've always made it a priority to listen to what my customers have to say and use that feedback to make informed decisions. Whether it's through surveys, reviews, or direct communication, understanding the needs and preferences of your customers is essential for long-term success.

Building Trust

Trust is the foundation of any successful business relationship. Customers need to trust that you will deliver on your promises and provide them with a positive experience. That's why I've always been transparent and honest in my dealings with customers. By building trust, you can create a loyal customer base that will keep coming back for more.

The Hassle Score

As we look ahead, I'm excited to introduce you to the Hassle Score, a unique assessment tool that I've developed based on my decades of experience in business. The Hassle Score is designed to help you understand, assess, and fix hassles in your business. By identifying and addressing customer hassles, you can improve the customer experience, increase loyalty, and drive profitability.

Removing hassles for customers is essential for long-term success in business. By understanding what buyers are seeking and constantly improving your offerings, you can create a loyal customer base that will drive profitability and growth. Embrace change, listen to your customers, and use tools like the Hassle Score to simplify the customer experience and set your business up for success.

In the chapters that follow, we'll delve deeper into the Hassle Score and explore practical strategies for simplifying the customer experience in your own business.

CHAPTER 1

Hassle Score Survey (HSS) & Key Performance Indicators

Customer feedback is gold. In my four decades of business, there is no more important metric to measure than how customers feel about your product and service.

In our current age of information, data is king. Understanding how prospective, current, and former customers feel about your product and service is the new gold standard in a crowded marketplace. Having a tool that gives you the business intelligence to know where you excel, and areas of growth that can transform your customer experience will help in turning them from prospective to loyal customers poised for long-term engagement with your business. In the process, this will keep your current and future competitors at bay.

If you have been or currently are in business, you are likely familiar with the adage that you cannot manage what you cannot measure. There is some truth to that. I prefer to say you cannot improve what you cannot measure. Therefore, today's gold standard is continuous improvement.

The Hassle Score Survey (HSS) measures how well a seller can provide a "no hassle experience" in the following four categories: convenience, price, product experience, and trustworthiness. The HSS tool will provide invaluable data, *aka gold standard*, that clearly and directly describes the entire customer experience from first contact through product or service delivery. In short, the HSS provides a roadmap of the entire customer experience that leads you into the path of organic and sustainable growth, which is how you will differentiate from the competition. You will become the gold standard in your sector.

I have found that there is not enough data on customer activity. Mainly, I have seen deep analysis around revenue, but not nearly enough focus on *what* drives that revenue. I advocate that sellers should diligently measure the number of customers gained and lost. I also advocate they should measure and understand the reasons behind gaining and losing customers. *It's absolutely the core of the business.* The Hassle Score Survey is driven by the reality that revenue drives sellers and businesses and is derived from buyers or customers. It is essential to understand the dynamics of customer gains and losses, and what leads to organic growth by fully understanding buyer activity.

Beyond a Familiar Score

I am familiar with the Net Promoter Score (NPS) and many companies or sellers like to use this metric. In my opinion, it's better than nothing, but not by much. The NPS is based on a single question, one! It's like someone thought that it would be too much trouble to ask buyers or customers too many questions. They might not like so many, or they would be hard to answer, or even expensive in spending company resources.

The one question they ask is this. "On a scale of 0-10, how likely are you to recommend our product/service to a friend or colleague?" That's it. My assumption is that this question is asked of actual buyers and not prospective buyers or shoppers. By not including prospective buyers or shoppers, you lose the information that is most critical. "Why didn't you buy from us?"

Back to NPS, there are three categories that answers are grouped into. For those that provide the highest score, either a 9 or 10, those are lumped into Promoters. Answers between 7-8, called Passives, are tossed out and answers between a score of 0-6 are categorized into a Detractors category. Basically, the Detractors are subtracted from the Promoters, and it results in a score. While better than nothing, the NPS is not a gold standard for assessing buyer activity.

I believe that sellers provide too many hassles to buyers and thus, for every hassle, the buyer is less likely to pull the trigger and make the purchase. I maintain that we buyers want and seek convenience in both the buying process and the product itself. We want competitive prices that are transparent, or we know what we are spending in total before we purchase. We want a fantastic product experience whether we are buying a service or merchandise, and we want to trust the seller. If we can get all four major categories, we are left with few excuses not to buy from a seller. If the seller cannot provide us all four of our major needs, we have excuses to buy elsewhere.

The Gold Standard - The Hassle Score Survey

The Hassle Score Survey is an adaptable tool that allows you the flexibility to either gain an overall summary score, or break it down in each category of Convenience, Price, Product Experience and Trust. This flexibility opens possibilities for more useful Key Performance Indicators (KPIs) to focus on areas of strength, and more importantly, where you need to improve your business. What is necessary to capture this information and key metric is talking to both prospects and buyers. You can execute this in a variety of ways, but a survey will work very well.

Using questions generated from the HSS will provide a better method for salespeople to ask and record. Empowering your sellers with this data gathering engine will improve their overall effectiveness and lead to greater revenue generation. Any seller will learn so much and be able to prioritize what they must work on to close more sales and eliminate more buyer excuses.

Prospecting for gold is difficult, but ultimately worth it. This is why the word "Eureka" has lived so long in our collective memories. While comprehensive and detailed, the Hassle Score Survey will be worth it.

Face it, when the buying process isn't convenient, it's a hassle for us buyers. When pricing isn't competitive, and we must shop around, it's a hassle. It's definitely a hassle when pricing isn't transparent, and you don't know the full cost before you pull the buying trigger. It's a hassle when the product experience isn't ideal. When we are treated rudely, when the product isn't up to par, when we can't ask questions, when the seller isn't living up to the brand promise, etc. It's all a hassle. When there isn't a guarantee, or they lie about the guarantee, there are no easy return processes or an ability to get a quick refund, those cause hassles. These questions are all designed to ensure sellers are minimizing or eliminating our hassle. Otherwise known as eliminating our buyer excuses.

Some of you might be jumping to the conclusion that collecting this type of information is going to be a hassle in and of itself. Customers and prospective buyers don't like surveys. They aren't willing to give this kind of information.

From my experience, I have learned that buyers and prospective buyers want to be heard and they want to help lower the hassle they go through. If they can help educate sellers into lowering the hassle score, they ultimately will be happier. Even if it does cost money to collect this information, it's far more valuable than nearly any other kind of data a seller regularly collects and measures. As a seller, knowing why you aren't closing and selling more business is invaluable and even if you must find creative ways to collect it, do it.

Competitors are lurking around every corner. The best competitors are going to figure out your weaknesses before you do and leverage those. If you are missing sales, you should know exactly why and then determine the best course of action to take to correct. Hearing directly from buyers and converting to a measurable and actionable KPI or set of KPIs will provide you a scientific method to plug your holes. All too often, sellers huddle and decide on internal thoughts, not science. Don't be one of them.

Finally, developing a KPI like the hassle score will support a culture of constant improvement and sustaining continuous improvement for each category and sub-category. We will be addressing the importance of having a culture of continuous improvement and having everyone involved in changes that will keep competitors at bay and eliminate the buyer excuses.

Hassle Ratio Survey

Survey Participants Instructions:
Please rate the following experiences of buying or considering our product/service compared to alternative options.

Convenience

1. When considering or purchasing from us, how do you rate the **time** it takes to make your purchase compared to alternatives?
   - ☐ Extreme
   - ☐ Significant
   - ☐ Average
   - ☐ Somewhat
   - ☐ Minimal
   - ☐ N/A

2. When considering or purchasing from us, how do you rate the **effort** it takes to make your purchase compared to alternatives?
   - ☐ Extreme
   - ☐ Significant
   - ☐ Average
   - ☐ Somewhat

- ☐ Minimal
- ☐ N/A

3. When considering or purchasing from us, how do you rate how **difficult** it is to make your purchase compared to alternatives?
   - ☐ Extreme
   - ☐ Significant
   - ☐ Average
   - ☐ Somewhat
   - ☐ Minimal
   - ☐ N/A

4. When considering or purchasing from us, how do you rate how **complex** it is to make your purchase compared to alternatives?
   - ☐ Extreme
   - ☐ Significant
   - ☐ Average
   - ☐ Somewhat
   - ☐ Minimal
   - ☐ N/A

Pricing

1. When considering or purchasing from us, how **competitive** is our pricing compared to alternatives?
   - ☐ Very Poor
   - ☐ Poor
   - ☐ Average
   - ☐ Good
   - ☐ Excellent
   - ☐ N/A

2. When considering or purchasing from us, how **transparent** is our pricing (knowing the exact cost before purchasing) compared to alternatives?
   - ☐ Very Poor
   - ☐ Poor
   - ☐ Average
   - ☐ Good
   - ☐ Excellent
   - ☐ N/A

Product Experience

1. When considering or purchasing from us, how **prompt are our responses** to your inquiries compared to alternatives?
   - ☐ Very Poor
   - ☐ Poor
   - ☐ Average
   - ☐ Good
   - ☐ Excellent
   - ☐ N/A

2. When considering or purchasing from us, how **clear is our communication** relative to our product and features compared to alternatives?
   - ☐ Very Poor
   - ☐ Poor
   - ☐ Average
   - ☐ Good
   - ☐ Excellent
   - ☐ N/A

3. When considering or purchasing from us, how **knowledgeable** is our staff compared to alternatives?
   - ☐ Very Poor
   - ☐ Poor
   - ☐ Average
   - ☐ Good
   - ☐ Excellent
   - ☐ N/A

4. When considering or purchasing from us, how would you rate your ability to **ask questions** before purchasing compared to alternatives?
   - ☐ Very Poor
   - ☐ Poor
   - ☐ Average
   - ☐ Good
   - ☐ Excellent
   - ☐ N/A

5. When considering or purchasing from us, how well do we **make you feel appreciated and special** compared to alternatives?
   - ☐ Very Poor
   - ☐ Poor
   - ☐ Average
   - ☐ Good
   - ☐ Excellent
   - ☐ N/A

6. When considering or purchasing from us, how would you rate our **professional appearance** compared to alternatives?
   - ☐ Very Poor
   - ☐ Poor
   - ☐ Average
   - ☐ Good
   - ☐ Excellent
   - ☐ N/A

7. When considering or purchasing from us, how would you rate our **cleanliness** compared to alternatives?
   - ☐ Very Poor
   - ☐ Poor
   - ☐ Average
   - ☐ Good
   - ☐ Excellent
   - ☐ N/A

8. When considering or purchasing from us, how would you rate our ability to **fulfill our brand promise** compared to alternatives?
   - ☐ Very Poor
   - ☐ Poor
   - ☐ Average
   - ☐ Good
   - ☐ Excellent
   - ☐ N/A

9. When considering or purchasing from us, how would you rate our ability to **pro-actively communicate and provide customer support** compared to alternatives?
    - ☐ Very Poor
    - ☐ Poor
    - ☐ Average
    - ☐ Good
    - ☐ Excellent
    - ☐ N/A

10. When considering or purchasing from us, how do you rate our **ability to personalize your experience** compared to alternatives?
    - ☐ Very Poor
    - ☐ Poor
    - ☐ Average
    - ☐ Good
    - ☐ Excellent
    - ☐ N/A

11. When considering or purchasing from us, how flexible are our **payment options** compared to alternatives?
    - ☐ Very Poor
    - ☐ Poor
    - ☐ Average
    - ☐ Good
    - ☐ Excellent
    - ☐ N/A

Merchandise Experience (If Applicable)

1. When considering or purchasing from us, how would you rate our product **durability** compared to alternatives?
   - ☐ Very Poor
   - ☐ Poor
   - ☐ Average
   - ☐ Good
   - ☐ Excellent
   - ☐ N/A

2. When considering or purchasing from us, how would you rate the **time** it takes to use our products compared to alternatives?
   - ☐ Very Poor
   - ☐ Poor
   - ☐ Average
   - ☐ Good
   - ☐ Excellent
   - ☐ N/A

3. When considering or purchasing from us, how would you rate the **effort** it takes to use our products compared to alternatives?
   - ☐ Very Poor
   - ☐ Poor
   - ☐ Average
   - ☐ Good
   - ☐ Excellent
   - ☐ N/A

4. When considering or purchasing from us, how would you rate the **simplicity** of our products compared to alternatives?
   - ☐ Very Poor
   - ☐ Poor
   - ☐ Average
   - ☐ Good
   - ☐ Excellent
   - ☐ N/A

5. When considering or purchasing from us, how would you rate the **ease** of our products compared to alternatives?
   - ☐ Very Poor
   - ☐ Poor
   - ☐ Average
   - ☐ Good
   - ☐ Excellent
   - ☐ N/A

6. When considering or purchasing from us, how would you rate the **stylishness** of our products compared to alternatives?
   - ☐ Very Poor
   - ☐ Poor
   - ☐ Average
   - ☐ Good
   - ☐ Excellent
   - ☐ N/A

7. When considering or purchasing from us, how would you rate the **quality of product packaging** compared to alternatives?
   - ☐ Very Poor
   - ☐ Poor
   - ☐ Average
   - ☐ Good
   - ☐ Excellent
   - ☐ N/A

8. When considering or purchasing from us, how would you rate the **quality of materials** compared to alternatives?
   - ☐ Very Poor
   - ☐ Poor
   - ☐ Average
   - ☐ Good
   - ☐ Excellent
   - ☐ N/A

9. When considering or purchasing from us, how would you rate the **efficiency of products** compared to alternatives?
   - ☐ Very Poor
   - ☐ Poor
   - ☐ Average
   - ☐ Good
   - ☐ Excellent
   - ☐ N/A

10. When considering or purchasing from us, how would you rate the **innovative features of products** compared to alternatives?
    - ☐ Very Poor
    - ☐ Poor
    - ☐ Average
    - ☐ Good
    - ☐ Excellent
    - ☐ N/A

11. When considering or purchasing from us, how would you rate the **safety of products** compared to alternatives?
    - ☐ Very Poor
    - ☐ Poor
    - ☐ Average
    - ☐ Good
    - ☐ Excellent
    - ☐ N/A

12. When considering or purchasing from us, how would you rate the **customizable options** compared to alternatives?
    - ☐ Very Poor
    - ☐ Poor
    - ☐ Average
    - ☐ Good
    - ☐ Excellent

13. When considering or purchasing from us, how would you rate the **maintenance required** of our products compared to alternatives?
    - ☐ Very Poor
    - ☐ Poor
    - ☐ Average
    - ☐ Good
    - ☐ Excellent
    - ☐ N/A

Trust

1. When considering or purchasing from us, how would you rate the **guarantees** provided compared to alternatives?
    - ☐ Very Poor
    - ☐ Poor
    - ☐ Average
    - ☐ Good
    - ☐ Excellent
    - ☐ N/A

2. When considering purchasing from us, how would you rate our **return proces**s compared to alternatives?
    - ☐ Very Poor
    - ☐ Poor
    - ☐ Average
    - ☐ Good
    - ☐ Excellent
    - ☐ N/A

3. When considering purchasing from us, how would you rate our **refund policies** compared to alternatives?
   - ☐ Very Poor
   - ☐ Poor
   - ☐ Average
   - ☐ Good
   - ☐ Excellent
   - ☐ N/A

4. When considering purchasing from us, how would you rate our **testimonials** compared to alternatives?
   - ☐ Very Poor
   - ☐ Poor
   - ☐ Average
   - ☐ Good
   - ☐ Excellent
   - ☐ N/A

5. When considering purchasing from us, how would you rate our **references** compared to alternatives?
   - ☐ Very Poor
   - ☐ Poor
   - ☐ Average
   - ☐ Good
   - ☐ Excellent
   - ☐ N/A

6. When considering purchasing from us, how would you rate our **warranties** compared to alternatives?
   - ☐ Very Poor
   - ☐ Poor
   - ☐ Average
   - ☐ Good
   - ☐ Excellent
   - ☐ N/A

7. When considering purchasing from us, how would you rate our **ability to deliver our brand promise** compared to alternatives?
   - ☐ Very Poor
   - ☐ Poor
   - ☐ Average
   - ☐ Good
   - ☐ Excellent
   - ☐ N/A

Instructions for calculating Hassle Ratio for sellers.

1. For each major category (convenience, price, product experience, trust), calculate the total possible points. Each question can carry up to 5 points, if "N/A" was not selected for a maximum of 10 to 55 in the different sections. If "N/A" was selected, do not count that question toward the total points.
2. Count the total points for each question, based on the answer. The lowest possible score is 1 and highest is 5. The answer "very poor" or equivalent will equal "1" and Excellent or equivalent will equal "5".

Example: When considering purchasing from us, how would you rate our **ability to deliver our brand promise** compared to alternatives?

- ☐ Very Poor (1)
- ☐ Poor (2)
- ☐ Average (3)
- ☐ Good (4)
- ☐ Excellent (5)
- ☐ N/A

3. Divide the total possible points by the actual scores recorded to determine the hassle ratio for each question, each category and overall. Subtract the results from 100% to determine the hassle ratio.

CHAPTER 2

The Four Criteria of Buyers

Starting my entrepreneurial journey as a young newspaper delivery boy, I quickly learned the value of hard work and determination. This early exposure to business sparked my curiosity and set me on the path towards future ventures. As I entered college at Springfield's Southwest Missouri State University, I saw an opportunity to turn my passion for business into a reality.

Inspired by the popularity of cashew chicken in Springfield, my college roommate and I decided to bring this beloved dish to our hometown of Rolla, Missouri. We secured the recipe and leased a former A&W root beer drive-in, eager to introduce our community to this delicious cuisine. However, despite our enthusiasm, we soon encountered challenges that tested our resolve.

While the food we served was excellent and delivered quickly, the limited space in our restaurant hindered the overall product experience. With only three tables available, our establishment primarily became a carry-out option, missing out on the opportunity to provide a comfortable dining atmosphere. This oversight proved to be detrimental to our success, and our business ultimately faltered.

Reflecting on this experience, I realized the importance of four core principles: convenience, price, product experience, and trust. Had we prioritized these principles from the outset, our venture might have had a different outcome. Convenience would have ensured that our customers could easily access our restaurant, while competitive pricing would have attracted budget-conscious diners.

Furthermore, prioritizing the product experience would have allowed us to create a welcoming environment for our customers, encouraging them to dine in rather than opt for carry-out. Finally, building trust with our customers through transparent business practices would have fostered loyalty and repeat business. These principles, when applied effectively, can make all the difference in a business venture's success.

As I look back on this early failure, I am grateful for the lessons. Today, I carry with me this experiential wisdom. With these four principles as a guide, I am confident in my ability to navigate the challenges of entrepreneurship and achieve success.

1. **Convenience**: Isn't it remarkable how a simple concept like convenience can significantly drive customer loyalty? Think of how Amazon revolutionized shopping with one-click purchases. How Jiffy Lube capitalized on 15–20-minute oil changes. How drivers will pay for the "Express Lane" to reach their destination faster. This book will explore how to embed convenience into every facet of your business, turning potential customers into loyal advocates.

   Convenience is anything that makes our lives easier, simpler, or less time-consuming. Convenience can include many things, such as tools, services, or products that allow us to complete tasks more efficiently. For instance, convenience stores are designed to provide quick and easy access to essential items, but convenience goes beyond that.

   If you've ever flown, you know that flying is often the preferred mode of transportation, even though it's only sometimes the most cost-effective or time-efficient option. So why do we choose to fly? The answer is simple: it saves us significant time and effort. When we opt to fly, we prioritize our convenience over other factors. However, trade-offs undoubtedly come with this decision, such as increased costs or potential delays.

In summary, convenience is a crucial consideration for many of the choices we make in our daily lives, and understanding its impact can help us make more informed decisions.

We will get into the details of convenience in the coming chapter, but you, as a business or seller, should start to think about what convenience you are offering your buyers. Your buyers must experience convenience in the buying process. The buyer's experience starts well before they make a purchase. Your product may not offer convenience, but you must be convenient when a buyer purchases. Ask yourself how much time it takes for a prospective buyer to buy from you, especially compared to other options. Ask yourself how much effort a buyer must exert to buy from you in relation to other options.

As a seller, how simple and easy is your offering? How simple and easy is it to buy from you compared to others? If your offerings are not convenient, buyers will find other options that are. If your product is offered at the same price as a competitor, but customers save time, effort, and gain ease from said competitor, you will lose opportunities to them.

2. **Price:** As a buyer, it's a natural human tendency to want a deal on price or discount with every purchase we make. We don't intentionally look to spend more than necessary for any purchase. This component yields more trade-off decisions than any other. We will pay more if it is more convenient than another option. We will pay more if the product experience surpasses another option and more if the product is more trustworthy than another option. It's how we buyers or humans roll.

   Price isn't just a number; it's a statement of value. How can you balance competitive pricing while maintaining quality and profitability? Here, we dissect pricing strategies that woo customers without eroding your bottom line.

Pricing must be competitive. If your pricing is not competitive, every seller must go to great lengths to justify the increase using other components, such as convenience, product experience, or trust. In this Internet-centric world, comparing pricing on products is far easier now than ever before. More and more, buyers detest the fine print and pricing surprises that are not shared upfront to help a buyer make a better buying decision.

When making a purchase, the price is often the crucial factor that buyers always consider.. Given the choice between two identical products or services, one on sale, we tend to opt for the discounted one. And when we do snag a bargain, we can't wait to share the news with our friends and family, boasting about our savings. For example, "I got $5000 off MSRP on my new car!" or "I bought my beautiful new yacht at a $100,000 discount because the dealer was overstocked." Even the wealthiest people enjoy a good deal, and no one likes to overpay.

In addition to a fair price, buyers also value pricing transparency. Unfortunately, sellers are not always completely transparent about their prices, particularly in the airline industry. You might have seen advertisements claiming that you can "fly for as low as $37," but the actual cost can quickly add up to over $150 once you factor in the seat assignment, carry-on luggage, soft drink, and early check-in fees. Deceptive advertising can leave buyers feeling cheated and deceived, believing that the ads are misleading.

As a seller, ask yourself how competitively you are priced compared to other options. How transparent is your pricing? In other words, are you luring buyers in with false advertising on the price only to have pricing gimmicks, fine print, or unknown rules? Buyers will look for other options if you aren't competitive and transparent with your pricing.

3. **Product experience**: With so many easy and fast options for buyers, product experience becomes essential. Product experience is when we get to use the durable good or merchandise or when the service we have purchased is delivered. Before we make a

purchase, be it merchandise or a service, we buyers perceive what we are buying and what should be delivered. When our perception of what we buy matches what is delivered, we are happy.

Product experience is critical to repeat customers and advocates for your brand. Product experience or delivering to what they advertise and what buyers expect is the one component sellers can control more than the others.

When purchasing a service or merchandise, the product experience holds significant importance. It's not just about buying a product, but how we are made to feel throughout the process. We expect to be treated as the only customer, with particular attention and care. After all, we are investing our hard-earned money and want to feel that it is being spent wisely.

The overall experience starts with the evaluation process until the product is utilized. We expect the experience to be exceptional every step of the way, not just during the evaluation of the product and vendor but also during delivery and recovery if necessary.

When we purchase merchandise, we expect high-quality, durable, efficient, and stylish products. But it's not just about the product itself. We expect the company to treat us as valued customers, even after we make a purchase. We want to feel important and unique, and being treated that way is crucial to our satisfaction.

As a seller, ask yourself, or better, ask your buyers about their product experience after engaging with you. Are you treating your buyers specially? Are you treating them as if they are your only customer? What are your competitors doing to make buyers feel special?

4. **Trust**: People buy from those they trust and products they know they can trust. When we buy, we perceive what we are buying and the associated product experience. The seller, too, projects an image of what they are selling and the experience they intend to provide. When those intersect and are delivered to expectation, we build trust in the seller.

In an age of information overload, trust is the currency of choice. How do you build and maintain it? This section provides strategies to establish unwavering customer trust and ensure long-term business relationships.

Whether buying products or services, we tend to choose those we can rely on. Our trust can be earned through previous experience with the product or service or through recommendations from people we know and trust. Additionally, we often look for testimonials, free or discounted trials, and flexible returns, warranties, and guarantees to help us feel more confident about our decisions.

The decision to buy is rarely based solely on trust. Many other factors, such as convenience, cost, and availability, come into play. Even if we like the people who work for a particular company, we may still choose to do business with a competitor if they offer a better deal or are more accessible.

How trustworthy are you as a seller? How are you attracting new buyers? Sure, existing or repeat buyers are hopefully loyal and continue to buy from you, but what are you offering to a prospect who knows nothing about you or your product? What are you doing to ensure they have confidence that what you say you are delivering is truly what they will get? What do you believe a buyer expects from you, so they don't feel buyer's remorse afterward?

As we embark on this journey to help you sustain growth, I challenge you to think: How do these four components exist in your current business model? Are they strong and well-balanced, or do they need reinforcement? By the end of this book, you'll have these answers and a clear path to making these pillars the foundation of your unstoppable growth. It's all about the buyer's experience from the first time they hear about you until well after the purchase.

As humans, we all share common ground regardless of our background or profession. We are all buyers. Whether purchasing goods and services for our personal use or buying as a business representative, we all are buying products. We are all in buying mode almost every day. My objective is to help you understand the intricacies of your purchasing decisions and provide a detailed understanding of the factors influencing your choices when selecting products or services. If you are a seller, fully understanding what buyers seek should enable you to cater to their needs and grow your business. Sounds simple! And it really is, but despite this fact, few follow the philosophy.

We make most of our buying decisions without thinking about the four criteria. We innately choose what we buy and from whom based on our ***personal value systems***. Our personal value systems are formed over our lifetimes that specify what is most important to us as people. When buying larger ticket items, our decision process becomes less innate and more deliberate, but we use the same four criteria.

For example, when my HVAC unit recently stopped working, I had to consider several factors before purchasing. Although my old unit was over 20 years old, and I had been expecting it to fail, I didn't want to replace a still-working unit. I wanted to squeeze every last second possible from the old one.

Given the time of year, convenience was essential, as we couldn't live without air conditioning for long. Equally important was finding someone I could trust, since I am not an expert on HVAC units. Lastly, I wanted a product experience, with the company delivering what they stated they could do and what I expected to receive as a buyer. As a buyer of a new HVAC, I wanted a competitively priced, reliable unit that would last me another 20 years.

Throughout this book, I will mention the term "product or products." For these, I am referring to both merchandise and services as those, to me, are a seller's products.

When we make our buying decisions for everyday items such as coffee, lunch, gasoline, and groceries, we innately make our decisions on what to buy and from whom to buy based on the four criteria. It becomes automatic.

As consumers, each of us bring unique values that influence our purchasing decisions. Some of us value our time over cost, some value cost over experience, and others value the product experience the most. Some buyers are guided more by trust than anything. We are all different. Personally, I would gladly spend more money if the product saved me time and effort, though there is a limit. Others will spend hours or days making a decision that will save $20. Furthermore, some will shop at Nordstrom or fly first class because they value the product experience over everything else.

To add another perspective, as buyers representing a company, we are guided by our *company's* value system. While the values may vary, the decision-making criteria remain the same.

I can tell you that Walmart places its highest emphasis on price. My first job was selling for Hunt Wesson Foods, and Bentonville, Arkansas, was in my sales territory. At the time, Walmart had nearly 200 stores, and I figured that selling 10 cases to each store would nearly satisfy my quota for ketchup that year. I made the appointment and, to my chagrin, left there without a deal as they had set the expectation that they would only buy it if I had nearly given them the ketchup. Naturally, Walmart sells Hunt's ketchup, but I understood it was all about price.

On the other hand, Northrup Gruman, a defense contractor, is all about quality and trust. Yes, price matters, but when assembling an aircraft, you can't buy solely on price as the aircraft must function each and every time it is supposed to.

Imagine you're walking down the street, and you come across two kids selling lemonade for the same exact price. While you might be more inclined to buy from the kid with the best smile or who makes you feel better, other factors could still influence your decision. If one kid is on one side of the street and the other is on the opposite side, you might buy from the one closer to you, even if you like the other kid more. Ultimately, with making purchasing decisions, there are many factors to consider, and different people will prioritize these factors differently based on their individual needs and preferences.

Many years ago, someone told me the expression, "All things being equal, people prefer to do business with people they like." While I believe this is a true statement, rarely are all things equal. Someone will be more convenient, priced more competitively or transparently, provide a better product experience, and be more trustworthy. Over the years, I can assure you I have lost a lot of business with people from the buyer's company that I liked, and I believed they liked me, but other reasons made them buy from another source. Those lessons are painful and can put a company out of business. If a seller thinks you are delivering an equal product and are simply trying to leverage relationships, that seller might be in for a rude awakening.

When "trust" is used instead of "like," the statement becomes more precise. If the buying process is straightforward and effortless, and the prices are both transparent and competitive, and if I am confident that the experience will be exceptional, I will purchase from the individual or company I have the most faith in.

When people make purchasing decisions, they usually have differing criteria. While some might prioritize finding the best value, the definition of "value" can vary. It is not necessarily synonymous with the lowest price, as that may result in inferior quality or performance or require more time and effort to take advantage of the lower price.

According to statistics, almost half of all new businesses fail within their first five years of operation, with approximately one-fifth failing within the first year alone. Although businesses fail for numerous reasons, no one intends to go out of business.

It's always fascinating to listen to entrepreneurs talk about their business ventures. They often discuss creating value, disrupting an industry, innovating, solving problems, or filling an unmet need. However, they seldom go beyond these general statements. They fail to provide details on what value they are creating, how companies are disrupting the industry, or what innovation they are introducing. They also need to specify the problem they are solving or the unmet need they are fulfilling.

When a company or business leader claims to offer value, it's like saying an athlete should excel in every sport. It's a broad statement without any substance. A more granular definition of value is needed to provide genuine value and create a successful strategy.

Value is a highly personal and somewhat elusive concept that varies greatly depending on an individual's beliefs, preferences, and priorities. However, we can still group these individual value systems into broad categories.

Regarding consumer goods and services, four main components tend to capture our attention and drive our purchasing decisions: convenience, deals, great product experiences, and trust. We, as consumers, tend to value these things the most.

Let's take a closer look at each of these categories. Convenience is all about making our lives easier and more efficient. Deals are about getting the most value for our money. Great product experiences are about enjoying the process and feeling satisfied with the result. Trust is about feeling secure and confident in our choices.

It's worth noting that the term "disruption" has become a buzzword in recent years, but it's not always the most accurate way to describe what's happening in the marketplace. Rather than simply shaking things up for the sake of change, true innovation is about finding new and better ways to offer value to customers. Innovation might involve making products and services more convenient, reducing costs, enhancing the product experience, or building trust.

Let's look at a few examples to understand the power of innovation. The cell phone, for instance, revolutionized how we communicate by allowing us to stay connected on the go. Artificial intelligence is another innovative technology changing how we work and play by making it easier to create content, analyze data, and make decisions.

Ultimately, the key to solving problems and meeting unmet needs lies in finding ways to enhance convenience, reduce costs, improve product experience, or build trust. By focusing on these core values, we can create products and services that truly resonate with customers and make a lasting impact.

Summary

With purchasing decisions, buyers sometimes use four specific criteria unconsciously. These criteria are determined by an individual's personal value system, which guides how they rate each. However, for professional buyers, the company's value system typically takes precedence in decision-making.

Value is a subjective concept that can mean different things to different people. Generally, though, it can be categorized into four main areas:

1. **Convenience**: People naturally seek ways to make their lives easier, including streamlining tasks associated with their purchases.
2. **Price**: Everyone wants to obtain the best possible value for their money and get the most for their investment.
3. **Product experience:** A positive experience with the product or service can significantly impact a buyer's perception of its overall value.
4. **Trust:** Buyers are more likely to engage with individuals or companies they can trust, as they have confidence in the quality of their offerings.

Imagine you are about to make a purchase. You have various options available, but you want to make the best decision based on your unique set of values. What factors do you consider before making your purchase? Do you know what drives your personal value system, and do you ever make changes to it? For instance, you may prioritize affordability over convenience when buying certain items but prefer the opposite for others. As a buyer, it's essential to understand what matters to you and why. Later, we will apply this same perspective to your role as a seller.

CHAPTER 3

Convenience

Venturing into home improvement projects used to be a daunting task for me, especially in navigating the vast aisles of Home Depot or Lowe's. I often found myself feeling lost and frustrated, desperately seeking assistance to locate the right product for my project. Struggling to find help resulted in wasted time and energy, hindering my progress, and leaving me overwhelmed by the store's sheer size.

Fortunately, my frustration was alleviated during a visit to my daughter's home in Atlanta, where my son-in-law introduced me to the Home Depot mobile app. This innovative tool revolutionized my shopping experience by providing detailed directions to the exact aisle and bin location of the items I needed. With just a few taps on my phone, I could efficiently navigate the store and locate my desired products without needing help.

The convenience offered by Home Depot's mobile app didn't stop there. With the option to pre-purchase items online, I could streamline my shopping experience even further. By paying with my credit card and selecting the "in-store pickup" option, I could avoid the hassle of searching for items in-store altogether. Instead, I simply arrived at the store, entered the provided code into the locker's computer, and retrieved my purchase from a secure locker in the entrance foyer.

This seamless process not only saved me valuable time but also allowed me to focus more effectively on my home renovation and repair projects. With the convenience of online ordering and in-store pickup, I no longer had to worry about getting lost in the aisles or waiting for assistance. Instead, I could confidently tackle my projects, knowing that the products I needed were readily available and easily accessible.

Reflecting on this, I am grateful for the convenience offered by modern technology and forward-thinking retailers like Home Depot. Their commitment to enhancing the customer experience through innovative solutions like the app has made a significant impact on my ability to complete home improvement projects efficiently and effectively. As I continue with future projects, I will undoubtedly rely on these convenient tools to simplify my shopping experience and maximize productivity.

In today's fast-paced world, convenience isn't just a luxury; it's a necessity. In this chapter, we delve into the heart and details of convenience, understanding its power to shape buying decisions and drive business success. Whether you're a budding entrepreneur or a seasoned CEO, mastering the art of convenience can catapult your business to new heights. Let's explore how time, effort, simplicity, and ease define the modern consumer's journey and how you can leverage them to create a compelling value proposition for your customers.

As buyers, our preferences are heavily influenced by convenience. We often select products or services that offer a higher degree of convenience than other options on the market. Convenience is a multi-dimensional concept that's hard to define in a single sentence. Instead of relying on standard definitions, let me share the various components that constitute convenience. From accessibility, ease of use, and time-saving features to location, availability, and delivery options, convenience encompasses many factors impacting our purchasing decisions.

Time

Time is a valuable and irreplaceable resource and can be even more valuable than money or physical assets! While opinions may vary on this, below are my researched conclusions as to why time is our most precious commodity.

Many people hire financial advisors to help manage their money. However, if you don't have one, you can calculate your net worth by adding up all your assets, including your homes, land, cars, boats, or other personal property checking and savings accounts, investments, and other possessions. Although this is an excellent approach to determining your worth, it's important to remember that financial advisors can predict how much money you will need or have at certain milestones in your life.

On the other hand, how much time we have left is unknown to anyone. Not knowing the value or amount of time should make us appreciate time even more. The Latin expression *carpe diem*, meaning "seize the day," reminds us that we should make the most of every moment. Some say, "I'm not here for a long time, but a good time." Life is short and unpredictable, and we must make the most of it. Ironically, people with too many assets tend to spend too much time thinking about what they will do with those assets when they run out of time!

I have vivid memories of my childhood in the late 50s and 60s when the aftermath of the Great Depression shaped my parents' spending habits. They only took out a mortgage; for everything else, they saved and paid in cash. If they wanted something outside of normal daily living, they planned their purchases and saved. The idea of being caught over-leveraged, or owing more than you had assets, was foreign to them. However, as the 70s rolled around, credit cards and credit terms became popular, and the culture of instant gratification began to take root. Today, people are more impatient than ever before. They want everything now, whether it's a new car, house, or an exotic vacation. More and more, we value time over other assets. The perceived lack of time drives us to be more aware of convenience and the drive to acquire assets and experiences faster.

Merchants have made the purchasing process more straightforward than ever before, and financing options have made it easier for people to buy whatever they desire, whenever they desire. Merchants include financing options in their advertising to overcome potential objections from buyers who don't have the cash to buy outright. It's no longer about saving for years to buy a big-ticket item; we can have it now and pay later.

The convenience of modern life has made it possible to save time in every aspect of our lives. We take the elevator instead of the stairs, fly instead of driving, and choose the fastest route to our destination. We buy groceries at our local stores and opt for quick-service providers like Jiffy Lube for our oil changes. We even pay extra for TSA Precheck or express lanes in metro areas to reach our destination faster.

Imagine shopping for groceries without the need to leave your house. With the rise of online shopping, stores like Walmart and services like Instacart have made it possible for customers to purchase items online and have them picked and bagged by employees. This service means avoiding the hassle of going into the store altogether, saving precious time. Companies like DoorDash, UberEATS, and Grubhub will save you time and effort by bringing your food order to you.

Time is a valuable commodity, and people are willing to pay for it. For example, air travel is a common choice for people who want to reach their destination quickly. There isn't anywhere you can go that you can't take a car, bus, train, or boat, yet we often opt for air travel as we save more time. And, when traveling by air, we evaluate nonstop flights and the cost compared to transferring to another airport because taking flights with connections costs us time!

At Disney, visitors can use the Lightning Lane and Genie+ options to bypass long lines and enjoy the rides faster than those with a standard pass. I think Disney has mastered figuring out how to extract dollars from its visitors by saving time or enhancing the product experience.

Consider how often you choose options that save you time when shopping for goods and services. Although it may involve trade-offs between price and product experience, time is often our most highly valued commodity. Intelligent merchants know this and offer time-saving options to attract customers.

My personal value system weighs heavily on the amount of time I must spend doing anything. I don't place a hard and fast dollar figure on my time, but I realize that time is money, and I don't want to waste it.

Effort

Humans opt for the path of least resistance, which can be interpreted as laziness. It's just a part of our nature. We prefer to conserve our energy whenever we can.

Let me illustrate this. At the airport, when you check in for your flight, the departure gate is usually located on the 2nd floor. There are escalators and stairs side by side, but how many people do you see taking the stairs? Maybe you will see a few fitness enthusiasts or travelers in a rush, while most prefer not to exert an effort and let the escalator do the work.

Humans often find ways to reduce the effort we put into everyday activities. We know that time and effort are closely related—usually, something that takes a lot of time also demands a lot of effort. And vice versa, something that takes a lot of effort will normally take time. There are a few exceptions to this rule, such as buying a car or a time-share, which can take up much time but require minimal effort.

We live in a world where automation has made our lives easier. Take, for example, the robotic vacuum cleaners that do the cleaning for us as programmed. Or the moving sidewalks that make it easy to get around airports and malls without breaking a sweat. Even if we must mow our lawns, we can use self-propelled or riding mowers instead of the traditional push mowers.

Washing our cars is another task where we opt for convenience. We can either do it in our driveways or visit a car wash to use the spray hoses ourselves. Washing the car at home requires one to get the hose unwound and get a bucket, a sponge or cloth, and maybe even a chamois or towel to the street or driveway. At a car wash, I can simply drive into a bay, throw a few quarters in the machine, and go away. Or, even better, I sit back in my car and go through the drive-through wash.

The ultimate convenience is the drive-through. It has become essential to many services, such as quick-serve restaurants, pharmacies, and banks. It requires minimum effort, as we can stay seated in our comfortable cars and get the same results. Interestingly, many fast-food restaurants are trying to optimize the drive-through experience. Chick-fil-A has added a door at the drive-through to make it more expedient for help to go back and forth into and out of the kitchen to fulfill orders before a car arrives at the window. Some locations now have dedicated lanes for those customers who order and pay via their mobile app.

But what if convenience comes at a cost? Often, we can park our cars, walk into the store, do our business, or order our food and leave the premises before it would have taken to go through the drive-through. Despite this fact, given an option, we almost always choose the option that requires less effort, provided there is no significant trade-off of time, experience, or cost.

In conclusion, we always look for ways to reduce our effort in everyday activities. We are grateful for the automation that has made our lives easier, and we often choose convenience over effort.

Simplicity

The formula for convenience is centered around simplicity. As humans, we prefer things that are uncomplicated and easy to understand. The more straightforward a task or process is, the better if it satisfies our requirements. For instance, the tax code is a complex subject many of us need help to comprehend.

For this reason, I employ a tax professional every year to calculate and submit my state and federal income tax returns, saving me both time and effort. The paramount reason for hiring a professional is to avoid making costly errors or overlooking exemptions, rules, calculations, and the like, which could result in paying more tax than necessary or missing out on a refund. Furthermore, I am adamant about avoiding an audit, which would take up too much of my time and effort. Wanting to do things right is a recurring theme in my life.

With the advent of technology, many innovations have been geared towards making our lives easier. Voice assistants like "Siri," "Alexa," and "Google Assistant" allow us to accomplish tasks with ease, such as inquiring about the temperature or adding an item to our shopping list, without having to go through the tedious process of looking it up or writing it down. Even the installation process of new software programs has become more straightforward, thanks to software companies' efforts to guide us through it. This way, we don't have to enlist the services of an IT expert, which is often a more expensive and time-consuming option. How many of us have tried to install a wireless network or set up an entertainment system and ended up calling the "Geek Squad" for help? I can relate to the frustration of reading complicated technical manuals, and I always favor simplicity whenever possible. Life is already complex enough, so any way to simplify it is always welcome.

Easy

In 2005, Staples launched an advertising campaign featuring the "Easy" button, which became a memorable symbol of their brand. The message was obvious: why take the difficult path when there is always a more straightforward alternative available? This idea has since been reflected in the popularity of YouTube, which is now widely recognized as the second most popular search engine. This is because of its delivery format, which answers questions through audio and video. Learning via video is much simpler and quicker than reading and comprehending text.

YouTube has an extensive library of content, particularly when learning new skills. It is a valuable resource for DIY enthusiasts who can find almost anything to tackle a project. However, hiring a professional is often the best option. For instance, I hired a carpenter when I needed a new fireplace mantle in my basement. Although I could have tried to do it myself, I knew the result would have been a different quality than a professional's. I knew doing it myself would require a lot of time and effort, especially because I don't have the natural or learned skills a tradesperson would have.

Even for less specialized work, hiring someone to do the job can be easier. For example, digging out old landscaping bushes and replacing them with new ones is physically demanding and requires effort and time. While it is possible to do it yourself, hiring someone to do it for you may be more practical. In such cases, the decision often comes down to a trade-off between the convenience of having someone else do the work and the cost of hiring them.

Nowadays, most service companies are focused on making things easier and faster for their customers. They strive to provide solutions that simplify tasks and save time, making it easier for people to get things done.

Summary

When making a purchasing decision, the convenience factor is often a critical aspect that buyers consider. The availability and accessibility of the product, as well as the ease of the purchasing process, can have a direct influence on how much weight we give to this factor. For smaller purchases, such as a cup of coffee, a tank of gasoline, or a lottery ticket, we tend to gravitate towards sellers that are conveniently located or readily accessible to us. However, for larger purchases, we adopt a more deliberate and analytical approach, considering all the criteria and making a conscious and informed decision based on all the factors, where the convenience factor may or may not be the primary consideration.

There are four significant components of convenience.

1. **Saving Time**: I believe that time is our most valuable resource, and I try not to waste it, nor do I appreciate it when someone else wastes my time. Often, the seller closest to me will help me save time. Personally, I believe time is our most precious commodity, often more precious than cash or hard assets.
2. **Saving Effort**: As humans, we tend to prefer the option that requires the least amount of effort. We're not necessarily lazy, but if given a choice, we tend to choose the path that involves less work.
3. **Simplicity**: As human beings, we often opt for the easier path to avoid unnecessary complexity. Given the choice, we prefer a more straightforward solution that requires less mental and physical effort. This tendency allows us to focus on more critical tasks and reduces stress.
4. **Easy**: We tend to opt for the easier path when we face a choice between something relatively simple and something challenging or complex. This inclination towards the path of least resistance can sometimes make us miss out on opportunities to grow and achieve our full potential.

The relationship between time and effort is inseparable. In most cases, tasks that require a considerable amount of time also demand a significant amount of effort. And tasks that require little time usually require a small amount of effort. Additionally, simplicity is often equated with ease of completion, as straightforward tasks are generally easier to accomplish than complex ones. As humans, we will most always opt for something being easy as opposed to hard.

CHAPTER 4

Price

Navigating the world of transportation and logistics, I encountered a pricing system that was anything but transparent. Pricing in the transportation industry, especially for less-than-truckload (LTL) shipments, was a labyrinth of obscure tariffs and convoluted discounts. As a shipper, deciphering these pricing structures required expensive technology platforms and complex calculations, leaving me frustrated and uncertain about the true cost of my shipments.

The lack of transparency in pricing not only created confusion but also gave rise to an entire industry dedicated to helping shippers navigate these complexities. While these third-party solutions provided some relief, they ultimately exacerbated the problem by further complicating an already convoluted pricing process. I realized the industry's reluctance to embrace transparent pricing mechanisms was costing both shippers and carriers dearly.

Reflecting on this experience, I realized the importance of price transparency in fostering trust and efficiency within the transportation industry. By providing upfront pricing information and simplifying pricing structures, carriers could streamline the shipping process and build stronger relationships with their customers. Transparency breeds trust, and in an industry as vital as transportation, trust is essential for long-term success.

As I delved deeper into the complexities of pricing in the transportation industry, I became increasingly passionate about advocating for change. I envisioned a future where pricing was clear, straightforward, and easily accessible to all stakeholders. By championing price transparency and pushing for industry-wide reforms, I sought to revolutionize the way business was conducted in the transportation sector.

With this vision in mind, I developed a mission to challenge the status quo and drive meaningful change within the industry. I collaborated with like-minded individuals and organizations to develop innovative solutions that would simplify pricing and empower shippers to make informed decisions. Together, we worked tirelessly to dismantle the barriers to price transparency and create a more equitable and efficient marketplace for all.

Through perseverance and determination, we saw the fruits of our labor as more carriers embraced transparent pricing practices. Shippers no longer had to rely on third-party solutions or navigate complex pricing structures; instead, they could access clear and upfront pricing information directly from the carriers themselves. This newfound transparency not only improved the customer experience but also strengthened trust and loyalty within the industry.

As I reflect on my journey in the transportation and logistics sector, I am proud of the strides we have made toward greater price transparency and accountability. By challenging the norms and advocating for change, we have transformed an antiquated and opaque pricing system into one that is transparent, accessible, and equitable for all. I remain committed to price transparency, leading to better buyer experience.

When purchasing products, we only sometimes opt for the lowest price option available. Whether you are a professional buyer or a consumer, this holds true. Don't get me wrong, price is extremely important, and every buyer wants to know the cost before they make a purchase, even more so than whether the product is convenient, provides a great product experience, or can be trusted. Both ordinary consumers buy everyday products such as gas, coffee, apples, and milk, and businesses make significant purchases for raw materials and services. If a low price were the only deciding factor, then stores such as ALDI would have a never-ending queue of customers, and stores such as Kroger's, Albertson's, and Hy-Vee would have gone out of business long ago. It would also mean that high-end brands such as Nordstrom, Range Rover, and North Face would have failed to establish themselves in the market, as they are not the low-price leaders in their respective categories. There would be no need for Hilton or Marriott hotels, as everyone would stay in a Motel 6.

Humans are naturally inclined to seek good deals when making purchases. Buying an item on sale and purchasing multiple items to get a discount is quite appealing. However, we don't base our buying decisions solely on price. We don't. We also consider the product experience, quality, convenience, time, effort required, and whether the product is trustworthy. We will opt for the lower-priced option if all these factors are equal. But rarely, if ever, are these other factors equal.

Many of us take pride in finding and buying great deals. Deals often spur us to buy, even when we don't need what we are buying right now. Even the wealthy appreciate a good bargain. They derive pleasure from acquiring a yacht their bank has foreclosed on or purchasing a helicopter on clearance. Some attribute their wealth to identifying and taking advantage of great opportunities. Several successful individuals bought something at the right time and turned it into something far more valuable. For example, when buildings, homes, or properties are foreclosed upon and sold, those with capital will buy them and ultimately turn them into investments or sell them at a profit. They don't necessarily need those buildings or properties but recognize an opportunity to profit from the investment.

However, we are generally not keen to admit when we knowingly overpay for something. We would rarely say, "I sure am glad I paid $1,000 over sticker for that new car." We all overpay at certain times; we prefer to keep that to ourselves. But, as a seller, your strategy best not be directed at finding suckers every now and then. It will not last.

Everyone loves a good deal, especially saving money on travel. But while we all like to get a great price on airline tickets, rental cars, or hotel rooms, we also expect the experience, convenience, and trust factors to be top-notch. Unfortunately, this is not always the case. With pricing, transparency is very important to buyers. Without knowing the exact cost of purchase before we buy, we can feel misled and cheated. We don't like unknown add-ons or the proverbial "fine print."

For example, many airlines are notorious for their non-transparent pricing strategies. They often charge hidden fees for flight changes, early boarding, premium seats, overweight baggage, and extra bags. These fees can surprise us and leave us feeling frustrated and ripped off. Many "fine print" and obscure rules irritate us because we often aren't educated or informed about them.

For example, when a flight is canceled or delayed due to weather conditions, we assume the airline will provide us with hotel accommodation and meals. However, most do not because their "rules" state that Acts of God, such as weather, are not their responsibility.

Another form of non-transparent pricing is associated with low-cost airlines. Take Allegiant as an example. They advertise one-way fares as low as $37, but that price excludes a seat, carry-on items, refreshments, and other amenities. You must pay extra even for a seat assignment of any kind. It is not a premium seat with additional legroom or an emergency row—any seat! The actual cost of your ticket can be much higher than what you initially expected.

These discount airlines also charge for baggage, whether you check it in or carry it on board. Beverages and snacks are not complimentary, and you may even have to pay for customer service. You can also buy priority boarding or opt for emergency aisle seats, which are more spacious and comfortable than regular seats. All these additional fees can quickly add up, and what you thought was a bargain $37 ticket will cost you $150 or more. It's the age-old trick to lure buyers in and make them believe they are getting a great deal, only to find the total price is not all that discounted or special.

It's not just low-cost airlines that have hidden fees. Even larger, more expensive airlines may have similar fees that passengers are unaware of before they fly.

Airlines aren't the only ones advertising one price, but they do not include all costs. Many hotels have added resort fees. They advertise rooms at $119 per night, but when you check in, you get assessed the resort fee, which supposedly covers internet, pool, and other amenities you might not even use. Some hotels have added this fee to combat third-party travel and booking sites where people often book solely on price. They discount those rooms slightly only to make up for it via their resort fee. It's not a transparent pricing practice, and many travelers don't like it, especially if it comes out of their pocket. Business travelers don't mind it as much as they can be reimbursed for whatever out-of-pocket costs they incur.

In the past, car rental companies used to charge based on the number of days, vehicle class, and the distance driven, which made it impossible to budget or anticipate the rental cost. Nowadays, car companies have introduced a new pricing method, charging flat daily rates based on the vehicle class, with unlimited miles. Although additional taxes may be added, this pricing method is still an improvement over the previous non-transparent one.

If you have ever booked a car through third-party sites like Travelocity or Expedia, you may have noticed that they advertise a price per day for the vehicle class and provide an estimated total price. However, it is not uncommon to see additional charges added up, which can increase the cost by 20-25%. Despite being an improvement from the past, the pricing method still needs to be more transparent.

Those of us who are old enough may remember landlines for telephone service. They used to charge by the minute, and while they told you that nights and weekends were cheaper, the phone bill was always a surprise.

Cell phone pricing has also evolved. In the past, users were charged a variable rate based on their usage, making it difficult to know how much they would spend on data usage. In addition to call use, cell phone providers charged fees based on data uploads and downloads, though consumers rarely knew exactly how many megabytes were being processed. The invoices for data use frustrate the buyers, specifically when they could not determine the amount of use. This lack of transparency made it challenging for users to monitor their expenses.

To put this into perspective, imagine hiring a painter to paint your house, and the painter quotes a rate of $100 per hour. If you ask, "How many hours will it take?" and the painter replies, "I'll let you know when I'm done," you would likely hesitate to hire them.

As buyers, we appreciate transparency and consistency in pricing. To make informed decisions, we want to know all the costs upfront, without any hidden fees or surprises. However, as we buyers become sellers, sometimes we forget how much we detest the non-transparency and opt for it in the name of profit or to lure buyers in under somewhat false pretenses.

Professional services such as attorneys, consultants, and accounting services often charge by the hour but don't provide a solid estimate of hours. Medical services are very poor at advising their pricing in advance. If you've ever had a hospital stay, you are mostly shocked by the cost and the line items assessed. Even with some hospitals being more transparent with pricing for services, it's often meaningless given the various insurance programs, Medicare and Medicaid programs, and how medical institutions negotiate with each other. Although we don't like these scenarios, we often have little other options for these types of services.

Other pricing schemes are unpopular with most buyers. For example, merchants that use coupons require customers to go through extra effort and spend valuable time cutting out the coupon, taking it to the store, and matching it up with the eligible items. Though the store doesn't always promote coupons, the manufacturer reimburses the store for the coupon amount. Unfortunately, the buyers must deal with the extra trouble. Yes, some don't mind the extra time and effort to cut and clip coupons, but imagine if they were afforded the same deal without going through the hassle? What is the point of a coupon? Why create a hassle for the buyer when we hate hassles?

Have you ever needed clarification about rebates, particularly regarding retailers? For those unfamiliar with it, rebates are a bit of an antiquated practice where the seller either refunds you off completed purchases or credits you toward future purchases. While purchasing a car sometimes entails a simple rebate process, redeeming a rebate offered by a retailer can often be cumbersome. Typically, it involves completing lengthy forms, submitting receipts, and mailing them without guarantee of receiving the rebate.

To add to this inconvenience, some retailers provide rebates in the shape of credit for future purchases, often marked up by 20-25%, effectively diminishing the value of the advertised rebate. Unsurprisingly, many people find this practice frustrating and often don't bother with it. Retailers use rebates to create goodwill and the illusion of a discount, hoping customers won't take the time and effort to redeem them. But is this the best way to go about it?

I find rebates that don't pay cash and substitute credits on their product work, much like golf tournaments. Most golf tournaments offer prizes in the form of gift certificates redeemable only in the pro shop.

However, those pro shop items are marked up 50%, so it's a double win for the golf shop.

Retailers should abandon this rebate system altogether and switch to more customer-friendly approaches!

A loyalty program is a much better alternative, and it can come in various forms. For instance, my local grocery store has a loyalty program that automatically applies discounts at checkout when I swipe my card. These programs are simple and easy to use, providing retailers with valuable information on their customers' buying habits. Retailers can better target marketing efforts by tracking customer behavior and creating a more personalized shopping experience.

In conclusion, rebates offered by retailers can be confusing and a hassle to redeem. A loyalty program is a much better approach, providing customers with discounts and convenience while providing retailers with invaluable customer data.

Transparent pricing is important for satisfying shoppers and, ultimately, for growth. Transparency provides for a much better product experience. Competitive pricing is equally important. So, how does a seller offer competitive pricing when they may not have the scale or purchasing power to compete with the "large sellers?"

How does a small seller compete on price with the large sellers with the buying power to purchase raw materials and goods at a lower cost than the small seller? I didn't say it would be easy. It may require a seller to take less margin than those with larger buying power and out-compete the big sellers on other criteria such as convenience, product experience, and trustworthiness. You must be enterprising and smart to find ways to reduce costs. Perhaps you find ways to cut overhead costs or live on lower margins and compete.

I realize that not everyone can compete and be competitive on price for many reasons. I didn't say this would be easy. It's not easy. Those sellers that cannot be competitively priced will likely have to figure out various ways to cut costs. Perhaps using new technology, becoming more productive, or sourcing differently. It will take a lot of effort, but it's why everything I preach is not a one-time fix. It can happen over time, but you must know where you need and want to go and align the right team behind you to help solve it.

Let's be honest. Each one of us bought something with unknown hidden fees or extra charges. We were not happy when we found out about them, as they were in the contract or fine print. We feel cheated. You likely say, "If I had known about that charge, I would not have made that purchase or done something different.

So, if we have those feelings as buyers, why do sellers continue the practice? Yes, every seller must make a profit to continue in business. But why risk the wrath of a buyer who will react negatively by telling others and not buying from that seller again?

If you are a seller, be competitively priced and completely transparent for long-term growth. Yes, with the amount of competition every seller faces, it's hard to deliver profits. But, even if you cannot be competitive with your price, overdo it with convenience, product experience, and trust. Tricking a buyer with surcharges, surprises, and extra fees will not yield long-term growth.

Summary

When purchasing, each buyer has a distinct set of preferences and priorities that guide their decision-making process. It's only sometimes that the buyer will opt for the lowest-priced option available, as several factors contribute to their overall buying experience. For instance, these factors may include the ease and convenience of the shopping process, the quality and features of the product being considered, the reputation of the supplier, and most importantly, the level of trust the buyer has in the supplier. These elements combine to shape the buyer's perception and ultimately influence their purchase decision.

1. **Competitive Prices**: Everyone loves a good deal. It's a fact. We often feel proud when we find and take advantage of a great offer. On the other hand, when we end up paying more than we should or more than the "market" price for something, we don't feel good about it and don't usually boast about it to others.
2. **Lowest Price**: As buyers, we sometimes opt for the lowest price. A low price often comes with inevitable trade-offs. The product or service might be time-consuming, complex, or challenging. It might also provide a poor product experience or come from an untrustworthy source. If we always chose the lowest price option, many well-known companies, like Nordstrom, Kroger, and Marriott, wouldn't exist today.
3. **Price Transparency**: As buyers, we dislike surprises when it comes to pricing. Fine print and hidden charges are not appreciated and can make us angry while leaving a bad taste in our mouths for the product and seller. We prefer to clearly understand the total cost of any product before making a purchase decision.

In a highly competitive market, sellers must remain vigilant about the pricing of alternative products and competitors. To stay ahead of the game, sellers must constantly strive to reduce their costs while keeping their product quality intact. By doing so, they can offer their buyers the most cost-effective deals without compromising on the value of their products. Pricing transparency is something that sellers control more than they do the list price of their product. While buyers often detest hidden fees and surcharges, fine print, and rules, sellers often ignore buyer preferences and hide extra fees.

CHAPTER 5

The Product Experience

As someone deeply entrenched in the logistics business, I've discovered the importance of the product experience in fostering long-term customer loyalty. To excel in customer service, I adhere to a set of principles that prioritize personalized interactions and prompt responses to inquiries. Each prospect or customer is treated as if they are the only one, receiving tailored interactions that make them feel valued and appreciated. By demonstrating genuine appreciation for their business and promptly responding to inquiries, I aim to show customers that their time is respected, and their satisfaction is paramount.

Furthermore, maintaining a professional appearance and fulfilling brand promises are essential components of delivering an exceptional product experience. Clean facilities and a polished appearance reflect positively on the business and also instill confidence in your customers. By upholding our brand promise and delivering on our commitments, we reinforce trust and credibility, further enhancing the overall product experience.

Proactive communication plays a crucial role in guiding customers along their buying journey and ensuring they feel informed and supported every step of the way. By providing timely updates and keeping customers informed through any developments, we empower them to make knowledgeable decisions and feel confident in their choices.

Additionally, offering quick and thorough recovery in the event of any mistakes demonstrates our commitment to customer satisfaction and helps mitigate any potential negative impacts.

In my approach to customer service, I prioritize swift order fulfillment and proactive communication to enhance the product experience. By responding promptly to inquiries and fulfilling orders quickly, we show customers that their needs are our top priority and that we are dedicated to their satisfaction. Furthermore, offering quick and thorough recovery in case of any mistakes helps maintain trust and loyalty, turning potential setbacks into opportunities to strengthen customer relationships.

Ultimately, the product experience is about more than just the quality of the product itself—it's about the entire journey from initial interaction to final purchase. By prioritizing personalized interactions, prompt responses, and proactive communication, we can create a positive and memorable experience for every customer. This commitment to excellence not only fosters customer loyalty but also sets us apart in a competitive marketplace, ensuring continued success and growth for years to come.

Product experience might sound like a strange term, but it is the actual consumption or use of the product, be it merchandise or service. I want to distinguish it from the buying process, the price, and the trust factor. It's the meat of what the seller is selling and what a buyer buys.

In a world where businesses fiercely compete for attention and loyalty, exceptional product experience is a beacon of untapped potential. Now, we will delve into the transformative role of customer interactions in propelling organic business growth.

Imagine a scenario where every customer feels valued, heard, and appreciated, a scenario where your business isn't just a provider but a cherished partner in your customers' journey, a partner dedicated to solving each buyer's problem. This chapter is not just about theories;

it's about practical, actionable strategies that can turn ordinary customer experiences into extraordinary tales of satisfaction and loyalty, much like a captivating scene from a beloved film.

Let's embark on this journey to understand how mastering the art of customer experience can become your most powerful tool in achieving unstoppable organic growth.

As buyers, we all hope for a delightful product experience when we purchase a product or service. We look forward to the item or service we buy to provide a top-notch experience, and we expect the purchasing process to be uncomplicated, effortless, time-efficient, personable, and low-maintenance. Most buyers share this expectation, and it influences their purchasing decisions. I don't think anyone enters the buying process knowing they will be treated rudely or disrespected and willingly accept this treatment.

Consider the transformative power of product experience, as vividly depicted in the film *Pretty Woman*™. When Vivian, portrayed by Julia Roberts, first attempts to shop on Rodeo Drive, she encounters dismissive and prejudiced attitudes. The first store surmises that this potential buyer likely doesn't have the wherewithal to afford their products and basically dismisses her and is rude. This negative experience is all too common in the business world, where first impressions can make or break a customer relationship. Within the first 30 seconds to four minutes, a person will decide about an individual, situation, or product. While that can change over time, the first impression is very important when buyers are making decisions. Contrast this with her later experience, where she is treated with dignity and respect, leading to a successful and satisfying transaction.

Once the manager of the shop discusses with Richard Gere's character, Edward Lewis, and determines they are going to spend an obscene amount of money, the attitudes of everyone working in the store change, and they all flock to Vivian's every need.

Now, let's translate this into the business realm. Imagine your business is the boutique on Rodeo Drive. Every client, regardless of their background or how they present themselves, is your Vivian. How you treat them can turn a one-time buyer into a loyal advocate for your brand or, conversely, a detractor who diminishes your reputation. The choice is yours.

As a seller, do you treat everyone equally, or do you discriminate based on your perception of their purchasing power or estimated profit? You see, more than convenience and price, every seller controls the product experience more than any other factor a buyer uses to decide.

More importantly, every buyer can be a repeat buyer or, even better, an advocate and promoter for you. Yes, they can even be detractors. That is, if their buying experience isn't up to par or what they expect, they will happily tell many others how poorly they were treated. They may not tell everyone how good their experience is, but they will *definitely* share negative experiences with others. As a seller, it's up to you what kind of impression you leave on a buyer.

In the following sections, we'll explore actionable strategies to ensure your business consistently delivers exceptional customer experiences. We'll dive into real-world case studies where businesses, much like yours, have transformed their customer interactions into a powerful engine for organic growth. Prepare to unlock the untapped potential within your customer service approach, turning every interaction into an opportunity for growth and success.

As you step into the doorway of a business establishment, have you ever been greeted by a friendly and enthusiastic staff member who asks: "Hello, is this your first visit?" Whether you answer affirmatively or negatively, the staff member will say, "Welcome back or thank you for visiting" and "How can I help you?" Although slight, this gesture makes us feel good.

As customers, we all want to feel important and cherished. Our satisfaction is just as important, if not more important, than anyone else's. Therefore, we anticipate being treated with the same level of respect and attention as anyone else. We want our time, effort, and finances to be acknowledged and valued. Simple and kind gestures, such as a warm welcome, can make a significant difference in how we perceive a business.

When customers feel they were not given the service they expected, they are more likely to share their experiences, particularly negative ones, with others. With the advent of social media, a single dissatisfied customer can quickly spread negativity, which can present a challenge for merchants and sellers to overcome. While positive reviews and experiences are always welcome, unhappy customers may express dissatisfaction by sharing negative opinions about how poorly they were treated. It's very important to establish and then maintain a great reputation for exceptional product experience.

Customer service is a vital aspect of any business that sells products or services. The overall experience of buying or prospecting revolves around more than just the product; it also involves the entire process. A seamless, hassle-free, time-efficient, and personable buying process that prioritizes customer satisfaction is essential in delivering excellent customer service. If the product fails to offer a great experience, it can be considered a waste of time, money, and effort.

As a seller, it is critical to maintain a high level of service throughout the entire process. From the moment a potential buyer shows interest in the product until after the purchase, ensuring that the customer is happy with their purchase and feels valued and appreciated is vital. Such a positive experience can result in glowing reviews and recommendations to others, significantly boosting the seller's reputation.

For service-based companies, exceptional service is defined by specific factors, such as prompt response times, clear communication, knowledgeable staff, and going above and beyond to meet the customer's needs. These factors can make all the difference in creating a loyal customer base and driving business growth.

Companies that sell durable goods are expected to go beyond just selling the product. They are expected to provide a service that begins before the purchase and continues even after the sale. Customers want a great buying experience, and thus, specific attributes need to be delivered. These include the product's durability, convenience, stylish appearance, good packaging, personalized customer support, efficient service, and product quality.

When customers make a purchase, they hold certain expectations and ideas about the product or service they buy. These expectations are often based not just on the buyer's imagination but also on the seller's brand promise or the product's advertised features. However, it is essential to understand that buyer experience is what determines their perception of the seller's brand. Therefore, it is crucial to meet or exceed the customers' expectations.

Let me give you a common example. We all eat out. Sometimes, eating out can mean stopping at a quick service restaurant such as McDonald's or Chick-fil-A. Other times, we might choose a restaurant where we dine in, like Panera or Applebee's as examples. Then, there are restaurants where it's more of a destination and perhaps to celebrate special events, like a chop house or high-end seafood establishment. As buyers, we have different expectations of what the product experience should be like at each.

In every instance, our minimum expectation is that the food should be fresh and, if cooked, hot, and have a decent taste. We do not expect the person behind the McDonald's counter to say. "Hello, my name is Darren, and I am here to serve you today. What can we start you off with?" On the other hand, we don't expect that same person to say, "Hey dude," or "What are you having, bub?" or "What's up?"

When we dine at the expensive destination restaurant where we make a reservation, we expect the server to share their name and indulge in conversation, make food recommendations, and even ask if we are there for a special occasion.

We expect something in between for sit-in restaurants like Panera and Applebee's. Still, we set an expectation in our mind and want at least the minimum product experience that such an establishment should provide.

Now, let's be a bit outrageous. What if the person behind the McDonald's counter gave us their name? "My name is Jim, and welcome to McDonald's. Is this your first visit with us? How can I help you?" That is a better experience than we likely expected. I always get a kick out of workers at Chick-fil-A. When I tell them thank you, they respond with "My pleasure." They are trained to do that instead of replying with a thank you. For me, it is more meaningful, different, and I take note.

Also, Chick-fil-A uses packaging with foil on the interior of their sandwich bags to keep the food just a little bit warmer and a little bit longer. That packaging costs a bit more but makes the product experience better than most other quick-serve restaurants. Have you noticed the science behind their drive-through process? Yes, it is more convenient, but it is also a better experience!

If we are eating at a high-end restaurant, we expect the food to be delicious, fresh, and even served where one isn't embarrassed to take a photo of it! If we get cold food served an hour after we order, we aren't happy. That isn't the experience we perceived we should get.

We all have visited or talked to businesses as potential buyers, and we get the feeling that the seller is doing us a favor to take care of or help us. I've heard business owners state that their business would be a great one if not for having to deal with buyers. Those impressions reverberate throughout the organization. Those sellers who understand the lifetime value of a customer typically are better suited and do a better job of providing an incredible product experience throughout the buying process and product delivery.

A company doesn't need to make a brand promise to create expectations in the minds of buyers. Even if there is no brand promise, the customer expects a certain level of service and quality. Some companies may choose to offer low prices or fast service, but it does not mean that the customers will accept a subpar experience. Ultimately, it is the seller's responsibility to meet the expectations they create in the buyer's mind. A satisfying product experience is essential to create happy and loyal customers who will return for more business in the future.

I would be remiss if I didn't mention sellers who are virtual or actual monopolies. By monopolies, I mean utilities such as gas, electric, and water companies. Railroads qualify as monopolies. Concessions at ballparks, arenas, and other venues that won't allow you to bring in refreshments are monopolies. Any seller that limits the options of buyers is a monopoly, and they do not care much about your product experience. You have no option, and so they can provide a less-than-ideal experience at a higher price with less convenience.

Yes, some monopolies are regulated by governments to keep them from totally ignoring the buyer, but for the most part, they are only driven by one thing: profit. They will typically invest in technology, processes, and procedures to become more efficient because, ultimately, they can make more profit.

Since I spent my career in the transportation industry, I had a beloved colleague at one time claim that in his next life, he was going to come back as a railroad. It was a joke, but he was taking a shot at their high prices, high profits, and terrible service.

All too often, I find sellers who are completely price-driven, meaning they aim to satisfy buyers on low price alone and aren't at all concerned about the product experience. They use their pricing as a crutch or an excuse to provide a mediocre product experience. It doesn't always cost more to provide a great product experience.

However, there are times when a great product experience *does* cost more. If operating a physical store, there are costs to keeping it clean, well-organized, brightly lit, and non-cluttered.

Let's say you are driving down the interstate, and you pull off to fuel. Two gas stations are immediately off the exit and are easily accessible (convenient). Both gas prices are the same, and both have a convenience shop attached. However, one is dimly lit, with oil stains on the concrete, cluttered windows with lots of advertising, and hasn't been painted or refreshed for years. The other store is well-lit, organized, clean, and fresh. Most would have the impression that the cleaner, more well-lit, and organized gas station would produce the better product experience. The outside appearance of the gas station might indicate the bathroom's cleanliness, which might be the determining factor as to which store to stop and buy gas. Product experience can be the total differentiator.

Sellers can control the product experience more than they can the convenience and price. It doesn't cost extra to treat prospects and customers special. It's a mindset and discipline that can set a seller above and beyond their competitors.

Summary

As a buyer, we want the buying experience to be as good or better than the product we are buying. We set expectations in our minds for both the product we are purchasing and the buying process experience. If we have moderate expectations and the seller exceeds them, we will feel immensely satisfied. Conversely, if we have high expectations, mainly due to a "Brand Promise," and the seller fails to meet those expectations, we will feel disappointed, as though our confidence has been misplaced.

Treated Specially: We, as buyers, expect personalized attention and respect when spending our hard-earned money on any product or merchandise.

Response: Time is one of our most precious commodities, and we recognize its significance in our lives. Therefore, we require speedy and efficient assistance when seeking solutions or purchasing services. If there are any complications, we anticipate a quick and effective recovery process with clear and concise explanations.

Quality: As buyers, we always look for a shopping experience that is enriched by the presence of knowledgeable and experienced professionals. We want individuals who thoroughly understand their offered services or merchandise. When it comes to services, I expect to interact with experts who are dressed in a professional and brand-appropriate manner and act competently and courteously. For merchandise, I am always searching for stylish, efficient but also durable, and comfortable products.

As consumers, we have high expectations and demand quality from our products, regardless of their price. We want to feel confident that we are receiving the best possible value for our money and only want excuses that may be correlated with the price we paid if it was explicitly stated beforehand. Essentially, we want to get what we paid for, with no surprises or disappointments.

CHAPTER 6

Trust

I've learned the hard way that trust is not easily earned, but it can be quickly lost. Recently, I encountered a Facebook ad offering HOKA tennis shoes at a remarkably low price of $39.99 due to a supposed inventory overrun. Having previously purchased a pair of HOKAs through another channel, I was intrigued but skeptical. My suspicions were heightened by an experience where I never received an item ordered through a Facebook ad, and obtaining a refund was a hassle. Despite the tempting offer, I ultimately decided not to proceed, unwilling to risk another disappointing experience.

Similarly, I was drawn to a clever USB video camera advertised on YouTube, promising enhanced Zoom calls with its compact design and suction cup attachment. Intrigued by the concept, I made the purchase after researching and seeing positive testimonials. However, upon receiving the camera, I quickly discovered that its quality was subpar, and the suction cup failed to hold after just a few minutes of use. Despite the vendor's return policy, the process was overly complicated, leaving me with a product that didn't meet my expectations and a valuable lesson learned.

These experiences have made me more cautious when encountering products advertised on various platforms, especially if the payment method is limited or the return process is unclear. I've come to realize that trusting a product's promises blindly can lead to disappointment and frustration. Instead, I now scrutinize the return policies, considering factors such as administrative fees, return shipping costs, and the overall convenience of the process before making a purchase.

Even in the realm of vacation rentals, where my wife and I sought warmth in Florida during the winter months, trust can be elusive. Despite finding what seemed like the perfect rental in Naples at a below-market price, our excitement quickly turned to suspicion when we discovered that the property's address was fictitious. We had fallen victim to a scam, losing over $1000 in the process. This experience served as a harsh reminder that trust should never be taken for granted, even on reputable platforms.

Throughout my career, I've encountered various scams and schemes aimed at exploiting trust for personal gain. While I've been fortunate enough to avoid some, others have left a lasting impact on my perception of trust in business transactions. In today's interconnected world, where communication channels abound, it's essential to remain vigilant and skeptical, always questioning the legitimacy of offers and claims that seem too good to be true. Trust is a fragile commodity, easily shattered by deceit and manipulation, but with vigilance and discernment, it can be preserved and nurtured in the face of adversity.

The foundational element of any successful business relationship is trust. As business owners, CEOs, and entrepreneurs, understanding and harnessing the power of trust is not just about avoiding scams but building a brand that resonates with reliability, authenticity, and integrity. From personal anecdotes to proven strategies, this chapter will guide you in creating a trust-centric culture that nurtures current customer relationships and attracts new ones, fueling your journey toward organic growth.

In a previous chapter, I mentioned the age-old saying that people tend to do business with those they like, assuming that all other factors are equal. However, all other factors are rarely equal. They are never equal. Nevertheless, this statement still holds, especially if you replace "like" with "trust." Trust is one of the four critical elements buyers consider while deciding whom and what to purchase. As I pointed out, I have bought products I had no way to trust, and I ultimately regret those purchases. Not all purchases, but some.

But what does trust mean?

"Trust" is the belief that the product being marketed is what is being delivered. It is also a matter of what you believe in your internal value system. For example, if a plumbing company claims or promises a response within 24 hours, it should be provided. But can we trust that just because they claim it? The obvious answer is no.

It was the late 1990s, and I was searching for a brand-new car. Having owned multiple Honda vehicles, I was drawn to that brand based on its reliability. After all, I valued a car that wouldn't break down frequently, as repairs could be a hassle, and towing or finding a substitute vehicle would only add to the inconvenience. However, I couldn't help but notice the soaring price tags attached to Honda cars. They were asking for a premium price.

That's when I stumbled upon a relatively new car brand recently entering the US market—Kia Motors, a South Korean brand. I was hesitant about buying a car from a country that wasn't renowned for producing high-quality items, and it reminded me of the days when Japan had a similar reputation. I wondered if Kia was in the same category.

Curious, I visited a local Kia dealer who had just opened a shop. The sleek Kia Optima immediately caught my eye as I entered the showroom. It looked visually impressive, and its price tag was significantly lower than a comparable Honda's. However, Honda had a reputation for producing quality cars, and I had experienced it firsthand. I couldn't help but worry about Kia's quality in comparison.

After speaking with a sales representative, I felt much more comfortable purchasing a Kia. Their powertrain came with a ten-year,

100,000-mile warranty, and their bumper-to-bumper warranty was five years and 60,000 miles, or much better than their competitors. For Kia to grow its market share and become established, it needed to offer a better warranty and a very competitive price. Given the price and warranty circumstances, I purchased the car that was on display, which was a fantastic vehicle.

Years later, when I passed the car down to my daughter, I was even more pleased with my decision. She was in an accident, and the driver's side door had an airbag that was deployed, saving her from serious injury. From that moment on, I became a big fan of Kia. They offer competitive pricing and excellent quality, and I trust them completely. Part of trust is knowing the buyer is going to be safe. Safety is a HUGE factor, and it is important for any seller to promote that as part of the features.

Trust in the product or seller is essential for anyone wanting to grow their business. Once a buyer purchases from a seller, they understand the convenience of the product and buying experience, the price being competitive and transparent, the product experience, and whether they can trust the product and the seller.

But, before the purchase, how should sellers convey trust to buyers?

Referrals

When it comes to determining the trustworthiness of a product or company, word-of-mouth referrals play a crucial role. We often rely on the experiences of our friends and family to make informed decisions. Their feedback is invaluable in shaping our perception of a product or company and can significantly influence our purchasing decisions.

As a seller, nothing is better than having other buyers refer you to buyers they know. They will do this of their own volition so long as your product and their buying experience are convenient, saving them time and effort, and it's simple and easy. Smart sellers ask customers for referrals after the sale and through an ongoing infrequent campaign. Smart sellers also are always asking buyers to be referred.

Having a second home at a lake house has made it challenging to find good solid contractors, for example. Finding contractors from Google is a way to have someone who doesn't do a good job or doesn't even show up. Having a referral system there is very important, or anywhere with fewer options due to lower population.

References

Many people use social media or other networking platforms to ask for recommendations about a specific company or product. Online reviews and testimonials are like referrals, but we may not know the person who responds directly in this case. However, we often rely on their feedback to decide whether to trust the offering. Reading online comments or listening to those on social media can be better than randomly choosing a product with no information or background.

Sellers should be proactive by offering references. As a seller, ask your buyers if they are willing to be a reference. Satisfied buyers will normally be happy to do that if you don't overuse them or ask too often.

Testimonials

Sellers often showcase testimonials on their website or other marketing materials to give customers a positive reference for their product or service. Amazon is a prime example of a seller that openly uses customer ratings and reviews to provide insights into the product and service quality. Other review sites such as Yelp and Google are also reliable sources of information, especially when personal referrals are unavailable.

When it comes to advertising, brands have the option to choose celebrities or influencers as their spokespersons. Celebrities are typically paid to promote the brand, but their endorsements are trusted by the public, even if they don't use the product. On the other hand, influencers are regular people who have built up a large following on social media platforms such as YouTube. Although they may not be famous, they are sponsored by brands due to their significant number of followers. Influencers have a relatable persona, which makes it easy for their followers to trust their recommendations. However, the brand still pays for their endorsements.

Sellers should proactively solicit and share testimonials to be used on their website, as websites are the ante for any business and seller. Buyers are very likely to check out the seller's website. In addition to an awesome value proposition message, pushing testimonials is an important part of gaining trust from potential buyers.

Guarantees

A guarantee is an essential factor that can help us trust the seller and product, especially when we believe the guarantee is genuine. However, we tend to be skeptical of claims like "satisfaction guaranteed." What exactly does satisfaction mean in this context? Moreover, how easy or difficult is it to get a refund under such guarantees? For instance, the convenience store chain QuikTrip offers a guarantee on their gasoline. It is unlikely that many claims have been filed, but this guarantee helps create a positive impression of QuikTrip in consumers' minds.

Sellers who are new or launching a new product should consider offering a guarantee. Be aware that if the product doesn't meet expectations, the guarantee will be redeemed. But, if you, as a seller, want to grow beyond repeat customers, a guarantee that is simple and easy to file is an important method to earn the trust of a new buyer.

Warranties

A warranty is like a guarantee but usually comes in written form with a lot of legal language, exclusions, and fine print. Despite the complexity of most warranties, we still rely on them. However, simpler warranties are always better, and sellers must be careful of scammers who try to take advantage for no reason.

Sellers should offer warranties for merchandise or items, especially if the price point is significant at all. This is a form of guarantee but enables the seller to "fix" the item or replace it if needed. As a seller, be straightforward and state the warranty in simple and easy terms.

Easy Return Policies

In today's world of e-commerce and the internet, allowing customers to return items easily and quickly is crucial for building trust.

Amazon has set the standard and is leading the way. Customers can return anything they purchase, and Amazon has made it highly convenient for them by partnering with chains like UPS Stores and Kohl's, among others. The best part is that you don't have to package the item, and the process to complete a return is straightforward. There are even times when Amazon will tell you to keep the product. However, I'm sure Amazon monitors customer behavior to avoid giving away too many items for free. Amazon has recently changed its return policies on certain items, and now a nominal fee is charged.

There is an ongoing debate and a few changes to some e-commerce companies' return policies. Certain buyers are taking advantage of return policies and "overbuying" and returning what they don't like. This is a common practice for clothing and shoes. Buyers select multiple sizes and colors and then return what they don't want. Returns are very costly for any seller. Thus, sellers must balance easy returns with monitoring and regulating individual buyers.

Free or Discounted Trials

Many sellers use a popular method to gain buyer trust by offering a free product version or a free/discounted trial. This tactic is beneficial for attracting new customers. Numerous online applications, such as Open AI, LinkedIn, and Grammarly, have free versions that provide a basic level of service or functionality, with the option to upgrade later. Even new companies looking to capture market share offer incentives to potential buyers to try out their product and prove its worth.

Depending on your product, I highly encourage sellers to adopt this strategy. There is nothing like being able to "test drive" any product, be it software or otherwise, to determine if it is a fit. I caution sellers who do this and collect payment information only to force buyers who don't elect to continue to go through a hassle to terminate their engagement. Just recently, the state of New York sued Sirius XM for making it too difficult for users to opt out of their service. You don't want to go through that as a seller if you wish to maintain a great reputation and grow.

Awards

Companies often promote awards they have won to build trust with potential buyers. Some companies also use terms like "best" or "number one" in their taglines to imply superiority, even if no governing body regulates the use of such terms and no one sues them for it.

As a seller, you absolutely should promote any awards or recognition you receive. It's a different kind of testimonial but will build trust with buyers.

Years of Experience

Some companies promote the number of years they have been in business to indirectly tell potential buyers that others have trusted them for a long time.

If you are a seller who has been around for over ten years, I advocate marketing this. Over half of new businesses close within five years. A seller's ability to stay in business for at least ten years is a great testimonial that buyers purchase from you and helps build trust.

As a buyer, I cannot stress enough how important it is to trust the seller when purchasing. Trustworthiness can be established through various means, such as displaying customer reviews and ratings, providing detailed product descriptions, and offering a customer-friendly return policy. However, regardless of the methods used, we always tend to buy from those we trust over those we don't.

Given a choice, we are going to buy the product we trust, and we are going to buy from a seller we trust. The more we learn about the product or seller from someone we trust, the better. Peers and family are great sources, but sellers don't know who their peers are, so they must provide us with proof to start the trust-building process.

I believe that many scammers, who even accept credit cards where charges can be reversed, know that a certain percentage of buyers aren't going to track every purchase and go through the hassle of filing a dispute. Sellers need to make themselves trustworthy, make the return process simple and easy, and then promote how hassle-free it is.

Sellers also need to be creative and work hard to extend their trustworthiness without having buyers take advantage of them.

As we conclude Chapter 6, it's clear that trust is not just a nice-to-have; it's a must-have in today's competitive business landscape. The stories, strategies, and insights this chapter share are more than lessons; they are the building blocks for creating a trustworthy brand. Remember, in business, trust is the currency of growth. As you continue your journey, let the principles of trust guide your decisions, interactions, and innovations. Your commitment to trust is not just an investment in customer relationships; it's an investment in your business' sustainable, organic growth.

Summary

Shopping for a product can be a daunting task, especially when it comes to trusting the seller. Sellers establishing trust is critical to ensuring a satisfactory purchase experience. There are various methods to achieve this, including requesting referrals or references, confirming warranties or guarantees, and checking for a generous return policy, free trials, or discounts. Evaluating the seller's experience in the industry and any awards or recognition they may have received can also help build trust.

Advertising is a powerful tool that many sellers use to promote their brand and build trust with their prospective customers. A well-known spokesperson, such as an NBA or NFL star, can go a long way in gaining trust and credibility. Their endorsement can sway people's opinions and perceptions towards the brand. It's essential to take note of these marketing strategies to make informed decisions when shopping.

Our priorities regarding trust may differ. These are a few contributing factors:

1. **Peers and Family**: The people we trust the most are our peers and family. If they have positive or negative experiences with a particular product or brand, we are likely to believe them.
2. **Guarantees**: Legitimate guarantees provide customers with a sense of security and confidence in a product or brand's quality, setting them apart from those who do not offer such assurances.
3. **Warranties**: Purchasing a product can be a daunting task, especially when shopping online, where you can't physically touch or see the item. However, one thing that can ease a buyer's mind is a legitimate warranty that stands behind a product. When a seller is willing to take responsibility for their product, it not only shows their confidence but also instills trust in potential customers.

In addition to a warranty, another way a seller can establish trust with customers is by offering simple, easy, and free returns. A hassle-free returns policy gives buyers peace of mind, knowing that if the product doesn't meet their expectations, they have a straightforward process to return it. This policy is also like a warranty or guarantee, further solidifying the trust between the seller and the buyer.

In commerce, companies often use language that persuades us to trust them. They employ words like "number one," "first," or "best," which appear authoritative and convincing. However, these claims may be hollow, with no empirical evidence to support them.

Sellers frequently tout the number of years they have been in business to gain our confidence. While it may seem like a compelling argument, it does not necessarily guarantee trustworthiness.

Unfortunately, many sellers overlook the importance of promoting ways to build trust with their customers. Make sure to provide assurances that you can be relied upon to avoid losing potential business. Trust is a crucial component of any value proposition, and with it, companies may find it easier to succeed in today's competitive marketplace. Repeat customers build trust by their actual buying and product experience. The real key for sellers looking to grow is to enable buyer trust at the onset.

CHAPTER 7

Competitive Advantage

Recently, I was reflecting on my daughter's cross-country journey and her coach's sage advice. Upon further reflection, I can't help but draw parallels to the world of business. The African proverb he shared resonates deeply with me: every morning, both the antelope and the lion must run, driven by the instinctual pursuit of survival. Similarly, in the competitive landscape of today's business world, every company must continually strive to outrun its competitors or risk being left behind.

Viewing it from this lens, we can surmise that success often hinges on identifying and exploiting weaknesses in existing competitors. When evaluating business opportunities, I advise aspiring entrepreneurs to look for weaknesses in established businesses, such as products that are slow to engage with, require excessive effort from the buyer, lack simplicity, transparency, or competitive pricing, fail to deliver a delightful product experience, or are perceived as untrustworthy.

In essence, finding a business or industry with one or more weaknesses presents a unique opportunity to carve out a competitive advantage and succeed. Just as my daughter's coach instilled in her the importance of speed and agility in cross-country, I encourage entrepreneurs to leverage their strengths to capitalize on competitors' weaknesses. By addressing these weaknesses head-on, entrepreneurs can position their businesses as formidable contenders in the marketplace.

Moreover, identifying multiple weaknesses in existing competitors can provide an even greater advantage. Just as a lion targets the slowest antelope in the pack, entrepreneurs can target multiple weaknesses to gain a strategic edge over their rivals. This multifaceted approach not only increases the likelihood of success but also reduces the risk of being easily outpaced or outmaneuvered in the market.

In my own entrepreneurial endeavors, I've witnessed firsthand the power of leveraging competitors' weaknesses to drive success. By identifying areas where existing businesses fall short and offering innovative solutions to address these shortcomings, entrepreneurs can carve out a niche for themselves and establish a strong foothold in the market.

Ultimately, the key to success lies in understanding that competition is inevitable and constantly evolving. Just as the antelope and the lion must adapt to survive in the wild, entrepreneurs must adapt to thrive in the competitive landscape of the business world. By embracing agility, innovation, and a keen understanding of competitors' weaknesses, entrepreneurs can position themselves for long-term success and stay ahead of the pack.

I am old school. I grew up under the pretense that you needed a competitive advantage as a company. I adhere to the concept of competitive advantage; at least, it's better than no advantage, but it's far better to have more than one. In contemporary business, it's far too easy for one seller to knock out another if all the one seller has is a single competitive advantage.

This chapter explores the delicate dance of catering to diverse buyer preferences in a market where convenience, price, product experience, and trust are paramount. We delve into the everyday decisions that shape company success and discuss how understanding and aligning with your customers' needs can propel your business toward sustainable organic growth. Through practical examples and insightful analysis, this chapter will equip you, the business leader, with the knowledge to create a strategy that resonates with your market, setting your business apart in a competitive landscape.

As buyers, we all want convenience, which can be achieved by saving time and effort and making the buying process simple and easy. We also wish for competitive pricing and a great product experience from a trustworthy seller. In short, we want it all!

However, buyers don't have a say in how convenient the buying process is, how much time and effort it takes, how simple and easy the process is, or how suitable the product is. We also don't decide the pricing or the product experience. As buyers, we can only determine if we trust the product and the seller, but we need the seller's help to make that decision.

As a buyer, have you ever been in a situation where sellers fail to deliver on the four criteria you look for? Have you ever wondered why this happens? It's because sellers, who are also buyers, often give excuses that discourage us from buying from them. They tend to leave gaps that alternatives can quickly fill, giving us buyers reasons to look elsewhere. For example, conventional car dealers focus on product experience, giving prospective buyers many vehicle options. From trim packages to colors and beyond, they strive to meet the picky buyer's need with an available vehicle that meets their exact specifications. However, buying a car via a conventional dealer is rarely convenient and priced competitively. New options like Carvana are making great strides in changing the car buying experience and eliminating buying excuses, especially in the used car market.

However, some companies stand out as models for not giving buyers reasons to search for alternatives. In this book, I explore these companies and analyze their strategies. I also examine companies that must appeal to buyers' personal value systems and make trade-offs.

We all have our personal value systems that have been shaped by our experiences over time. For instance, some of us prioritize convenience over price and product experience, others value price over product experience or convenience. However, if the price is too high or the experience could be better, we might choose to forgo convenience.

Our personal value systems, as buyers, drive us behaviorally to sellers who best match that value system, be it convenience, price point, product experience, or trust.

When it comes to buying products, different people have different priorities. For some, the product's price is of utmost importance, even if it means spending more time and effort to make the purchase or if it compromises the overall experience. On the other hand, some people prioritize the product experience over price and convenience, or they may prioritize either.

As buyers, we all have our own trade-offs and limitations intrinsic to our personalities and likely developed over a long period. Similarly, sellers also face trade-offs regarding pricing and the overall experience of their product. For instance, they may have to offer a lower price or a better product experience.

To better understand this concept, let me give two examples of buyers and sellers facing these trade-offs.

I have a friend who resides in a small town with about 15,000 people. The town has two grocery stores in addition to Walmart. Each week, my friend spends over two hours meticulously clipping coupons, studying ads, and traveling to each store to buy groceries and supplies that will last for a week or even longer. Whenever there is a significant discount or sale for certain products at any store, my friend will buy supplies for a few months. Although I am still determining, she saves around $20-$25 each week as it is only for her and her husband.

My friend's personal value system prioritizes pricing and deals. She is not concerned about the time, effort, or convenience she spends while shopping if she secures the best deal possible for what she needs.

In contrast, I value my time much more than my money savings, and therefore, I prefer to shop at the closest grocery store, even if it means paying a little extra. While there, I take advantage of the store's promotions but don't go out of my way to do so. My time and effort are more valuable than the savings I would get from shopping around.

Jiffy Lube is a famous oil change company that offers quick and efficient services. They specialize in changing your car's oil in 15-20 minutes without requiring prior appointments. Car owners can arrive and have their cars serviced without long periods of waiting. On the other hand, Walmart also provides oil change services, but their process may take over an hour to complete. However, they offer their services at a significantly lower price than Jiffy Lube or other similar companies. Walmart is designed to cater to individuals who prioritize saving money over time.

Both companies target different segments of buyers who use their personal value systems to make purchasing decisions. Jiffy Lube caters to individuals who value their time and are willing to pay a higher price for quick and efficient service. Meanwhile, Walmart caters to individuals who prioritize saving money over time and are ready to wait a little longer for their car to be serviced.

As a seller, it can be challenging to meet all four criteria buyers use to make purchasing decisions. Different companies make trade-offs that require buyers to engage their personal value systems to make the most meaningful choices. Ultimately, it's up to the buyer to decide what they value most: convenience, price, product experience, or trust.

The book titled *The Discipline of Market Leaders™* (written by Michael Treacy and Fred Wiersema in 1995) provides valuable insights into the three value disciplines businesses can focus on to excel in the market: operational excellence, product leadership, and customer intimacy.

1. **Operational excellence** is all about streamlining operations and cost leadership. It entails delivering a combination of quality, price, and ease of purchase that no one else can

match. Companies like Walmart and Amazon, which rely on efficiency and economies of scale, are perfect examples of businesses that excel in this discipline.

2. **Product leadership**, conversely, centers around innovation and brand marketing. Companies focusing on this discipline aim to produce a continuous stream of state-of-the-art products or services. Apple is a prime example of a company that consistently innovates and has an unparalleled brand cachet.
3. **Customer intimacy** involves offering tailor-made solutions to clients. Companies that excel in this discipline have a deep understanding of their customer's needs and can tailor their products and services to meet those specific needs. Salesforce and Nordstrom are great examples of companies that deliver customized solutions and exceptional customer service.

It is essential to choose one discipline and excel at it, according to Treacy and Wiersma. They claim that attempting to be all things to all people would result in mediocrity across the board. The book also discusses the organizational implications of selecting a discipline. For instance, companies focused on operational excellence will likely have a different culture and internal processes than those prioritizing customer intimacy.

Deploying successful business strategies requires a comprehensive approach that involves aligning your organization's various facets, such as operations, culture, and resources, around your chosen discipline. This strategic orientation should guide your business decisions and growth, and more than a one-off effort is needed.

As the book suggests, when I started my first business, I only focused on gaining a single competitive advantage. Competitive advantages are built along the same criteria buyers use to make buying decisions. Companies can leverage convenience, price, product experience, and trust, but few leverage all four. Nowadays, companies should strive to excel in as many criteria as possible that buyers use to make decisions.

It gives buyers fewer excuses not to buy from a seller who offers more of the criteria.

Having a single competitive advantage in one area is like having a castle with only one wall. While you might be protected from attacks from one direction, you are vulnerable to attacks from the other three sides. Therefore, you must know your weaknesses because your good competitors will evaluate and attack them.

Bill Belichick, the infamous football coach of the New England Patriots, is well-known for identifying his opponents' weaknesses and exploiting them. A similar approach can be applied in the business world. If you focus on convenience as your primary competitive advantage, then adding lower pricing or better buyer experience should enable you to exploit the industry standard for advantage.

If you can find a business or industry with one or more weaknesses, it presents an opportunity to start a new business attacking the competitors' weakness and succeed. And if you can find more than one weakness, that's even better since it will make your new company less likely to be knocked off in short order.

Imagine you are an aspiring entrepreneur eager to embark on a new business venture. You have a passion for the type of work you do and are excited to get started. However, before you dive in headfirst, it's important to remember that passion alone won't guarantee success. You must meet the needs of your targeted buyers.

Just because you enjoy cooking doesn't mean that opening a restaurant is the right choice. Similarly, if you have a talent for woodworking, that doesn't necessarily mean that carpentry is the best business option. Also, having an affinity for working outdoors and with your hands doesn't mean starting a landscaping business will be a guaranteed success.

Instead, consider focusing on filling a market need. If there is a demand for a particular product or service that still needs to be met, that could be the perfect opportunity for you. By providing convenience, competitive pricing, and a great product experience, you can set yourself apart from competitors and build a loyal customer base. By understanding and addressing the market's needs, your chances of success can be significantly higher.

For years, the cab industry focused completely on one advantage to gain traction: convenience. For years, they succeeded as the best option to catch a ride between any two points in most cities. Cities like New York and Chicago had ample cabs roaming the streets of downtown, and you had to stand on the sidewalk and wave one down. However, you had no idea what the cost was going to be. It was very opaque. On the side of most cabs, they indicated the initial cost for the first one-quarter mile and then a rate per mile thereafter. However, you didn't know the distance, as when cabs were most popular, you couldn't access many mapping applications on your smart device.

Once you got in the cab, you sat behind a plexiglass barrier, spoke to the driver, who might or might not be fluent in English, and endured the ride on a very uncomfortable seat. Once you arrived at your destination, the meter indicated the total charge, and you paid either cash (cab drivers always asked for cash) or provided a credit card, which typically took several minutes to complete. But you arrived at your destination.

In 2009, Uber, a ride-sharing service, launched targeting the "black car" or limousine service. They determined that the market only offered very expensive limo services with minimal options. Their original idea was to have owners of fancy SUVs, or other large cars offer the service but at a lower cost and allow the car owner to operate their own businesses.

It didn't take long before Uber decided to expand and enable pretty much any car owner to leverage that ownership and begin offering rides. They already had a great application that operated from any mobile device and offered a larger variety of car types or rides.

Uber didn't just enter the market with a single competitive advantage. They blew away the cab industry with more convenience, lower pricing, a better product experience, and a service one could trust by seeing how others evaluated individual drivers.

Using a mobile app, a rider didn't have to wave down a cab or, even better, call some dispatcher asking to send a driver to their pickup location. A rider was notified of the driver, vehicle type, and when the car would be there for the pickup. Heck, they even enabled the rider to see the car traveling to the pickup point on a map interface. Once there, the rider got into a car, likely much like their own, without a plexiglass partition, comfortable seats, and no meter running. When you ordered the ride, the application told you exactly what the cost would be before you took the ride. No more opaque pricing.

The driver could be a neighbor, which was much like catching a ride with a friend. And, best of all, when you arrived at your destination, you exited the car and walked away. The application had your credit card and charged the fare. More convenient than a cab? Yes. More competitive than a cab ride? Bingo. More transparency on the cost before you purchase. Absolutely. Better product experience. You bet. A ride you could trust? Absolutely.

Uber has practically put the cab industry out of business. They deliver on all four of the criteria we buyers have when we select a product to buy. Had they only focused and delivered on one of the criteria, I'm certain some other ride-sharing company would have come along and knocked them off.

As a seller, the more ways you can fulfill the buyer's criteria, the better you can compete, thrive and grow.

Summary

When making a purchase, buyers typically have four primary considerations in mind. They are looking for convenience, which means saving time and effort, competitive and transparent pricing, great product experience, and trustworthiness from the seller. The buyers will most likely select the sellers who can fulfill most or all these criteria. If I can buy the same product from two vendors, both are priced equally, and one provides more convenience and a better product experience, I will buy from the one I benefit most.

You and I are buyers. We both make purchasing decisions nearly every day. Not all of us are sellers, however. Sellers need to understand exactly what goes through a buyer's head when making buying decisions. Sellers must put aside what is important to them and focus on what is important to the buyers.

Some sellers may emphasize convenience and price more, valuing the former over the latter. Others may prioritize low prices, even sacrificing some convenience or product experience. Finally, some sellers may focus on providing an exceptional product experience, even if it comes at a higher price, arguing that the benefits of such an experience justify the cost.

Too many sellers focus on leveraging a single competitive advantage, or only one of the criteria we buyers have when making purchase decisions. By only catering to one of the buyer's criteria, those sellers are far too much at risk from a competitor knocking them off and taking business away.

CHAPTER 8

The Formula for Dominant Companies

Background

We turn our attention to the concept of dominant companies—those that have mastered the art of fulfilling the four key buyer criteria: convenience, price, product experience, and trust. This chapter explores how these companies have set themselves apart in the market, drawing lessons from their strategies and buyer experiences that can be applied to any business. Through compelling examples, we will understand how aligning your business practices with these criteria can satisfy and delight your customers, paving the way for sustainable, organic growth.

I would like you to consider something right now. Other than something you have no options to buy elsewhere, such as utilities, think about the products (merchandise or service) you buy regularly. Do you purchase such merchandise or services from the same seller every time? You likely get into habits and form perspectives without thinking much about it. For lower-cost items, we make our buying decisions innately. Ask yourself why you buy what you buy and from whom you buy. I'm especially interested in you evaluating those everyday type purchases and not major items. Are you making your purchase decisions on one or more of the four major criteria? Is it convenient for you? Is the product priced competitively and transparently? Is the product experience what you expect? Do you trust the seller and product?

Not to digress, but almost every monopoly is regulated. Why? Because a governing body is trying to force these companies to deliver the same buyer criteria as if they had actual competition. Otherwise, like any good monopoly, they would provide a very inconvenient product at a very high and non-transparent price point while delivering a less-than-desirable product experience that no one could trust.

We typically buy small ticket items without thinking much about them. We make these buying decisions mostly based on how our personal value system is programmed. We only take a little time to think about where to buy our morning coffee on the way to work, or where we are going to buy gas. Innately, we think about what coffee or gas is most convenient, what price we are willing to pay, and what product experience we are seeking, be it hot and delicious coffee prepared precisely like we want it or clean and well-lit gas stations and what product and from what seller can I trust. We don't sit and think about it for 10 minutes; it's more of a natural reaction, and we often buy identical products from the same seller repeatedly because it checks all our criteria.

Each of us has a distinct personal value system influencing our buying decisions. It's doubtful that we apply the same unique value to all our purchases. The higher the value or price of our purchase, the more likely we are to stop and think about what criteria are more important to us and what options are available. Some of us are driven more by convenience, or a better product experience than convenience, trustworthiness more than price, price more than convenience, etc. We are all slightly different, and it varies by the expenditure.

Let's use this example. Imagine you are in the market for a new TV. Your current one is ten years old and starting to have issues. You likely don't run down to the closest store that offers TVs and buy the first one you see. You are likely to research various TV brands and models. You might be driven to save time and effort, so you visit a store with many options, and they will even deliver and install. That option is the most convenient. You might be driven by price point. Thus, you are evaluating all options online in addition to the websites of local stores, comparing brands, models, and pricing. Your personal value system might drive you to the best product experience. The brands and models with the highest picture quality have the most robust sound system and the thinnest screen and can connect to other devices like Siri or Google. You may also determine that you will only be satisfied with a certain brand and seller because they offer the best warranty, guarantee, and returns process. If you cannot purchase a TV that meets all four criteria, you will select the seller brand and model based on the values that most drive your buying decision.

When grocery shopping, I often buy products based on pricing, opting for the private-label or generic brand as they tend to be cheaper than the name brand. However, I make exceptions. When I tried the generic brand in the past, I found that its taste/quality could have been better with the name-brand alternative. For instance, I prefer Coca-Cola over privately labeled brands of soda. But I can't differentiate much between private-label potato chips and national brands, so I tend to buy the lower-priced ones.

Even among national and well-known brands, there can be differences. For instance, let's take toothpaste, which Crest and Colgate offer threaded and flip-top tubes. In this case, I prioritize convenience over the brand and go with the flip-top options as they're easier to open with one hand while holding a toothbrush in the other. While many don't think about it, your buying decisions boil down to the value you prioritize, but when more than one is offered, you likely lean toward that product.

You will start to identify your personal value system when you evaluate what and from whom you buy. You will understand the trade-off you are willing to make and likely to what extent that trade-off will be.

Now, think about products you won't buy. Why? Why do you purchase the coffee you buy? Is it the cheapest? Is it the best flavor? Is it always hot? Is it available very quickly when you need it? Is it from a seller you can trust to deliver to your specifications each time? Do you trust that it is going to be consistent? Something is driving those decisions. I doubt you buy your coffee from someone different all the time. You have decided what you value the most and innately make that decision.

You are also pondering how much of a difference it would make to trade off something. The larger the expenditure for your purchase, the more you will weigh the trade-off and attempt to meet as many benefits or buying criteria as possible. Remember, the TV you are buying will take you more time to make that decision than that cup of coffee you buy every day on the way to work.

Companies that Dominate

Now, I'm going to use coffee as an example. You have a seller located just a block away; they have a drive-through with double lanes, and the line moves quickly. One of the lines is reserved for customers who purchase from their mobile device application and prepay. It's easy as the app remembers what you order, so you click an icon or radio button, and it's ordered. Your payment information is recorded, so it's automatically charged. Oh, and their application has facial recognition capabilities to log in (saves time and effort and is simple and easy to log in). And you can order coffee precisely the way you like it, customized! This coffee shop has been around for 20 years and has been voted the best coffee in your community. It's priced like the coffee shop just a block further down the street. In other words, this coffee shop does not give you *any excuses* to buy coffee from anyone else.

You have no excuses to go elsewhere, and are now loyal. This seller dominates their market because they have mostly eliminated excuses buyers use to purchase elsewhere. Now, the coffee shop down the street has excellent coffee that is customized but only has one drive-through lane that is usually backed up, and they charge a premium for their coffee. You have an excuse for a couple of reasons to go elsewhere.

This seller has elected to cater to a segment of the market. Most sellers do this, and it won't put them out of business; it will simply shrink the number of buyers willing to engage. They likely also target the buyer segment in the market that matches their strength or product experience.

The first coffee shop is a dominant company. While they have great coffee, fast service, a loyalty program, and very convenient and fast service, they don't charge a premium. There is no trade-off.

In the last chapter, we described how Uber decimated the cab industry. Why? They delivered on all four of the criteria we buyers have. They minimize the excuses to buy from another source, like a cab.

Many of you likely have shopped or are Amazon customers with Prime membership. They are a dominant company and offer buyers fewer reasons or excuses to shop elsewhere.

Convenience: I read a post on Facebook a while back stating, "If I had to check myself out, I would simply stay home and order from Amazon." You can buy something with 1-click on Amazon. You can set up repeat purchases to save you even more effort for those items you often go through. Their search engine is impressive, and you will always get tons of choices. You can get a 2-day delivery service and the same or the next day, depending on your location and the merchandise you are buying.

Price: Amazon is a marketplace, and vendors or sellers must compete for market share by its very nature. Amazon also contracts with them, and the seller must promote the lowest pricing for another vendor's sales channel. The price is upfront and transparent, and if you are a Prime member, they offer free shipping.

Product Experience: When evaluating products, you can see what other buyers think about the product you are researching, and they don't hold back the negative comments either. They ensure that the ratings are provided by those who purchased the product to get accurate and meaningful feedback from potential buyers. They also provide a forum by which you can ask questions of other buyers or sellers on specifications, features, or general use.

Trust: Amazon set the bar for liberal returns. Yes, there is a time limit, but a buyer can decide to return any item with only a few questions asked and will get full credit. They have partnered with UPS Stores, Kohl's Department Stores, and others to process your return. You can bring back a return unpackaged, and so long as you have captured the barcode that authorizes the return, you walk hand over the return, they scan your barcode, and you are on your way. What? They save buyers time and effort on returns!

Are they the absolute cheapest for every product? No, they don't have to be, as they provide sufficient options to satisfy the buyer's needs; over the long haul, they give buyers fewer excuses to shop elsewhere than others offering merchandise. They also have a vast array of merchandise they offer.

For those old enough, go back to when Blockbuster was the number one source for non-network movies and TV series. Have you, like me, ever been disappointed on that Friday or Saturday night, walking into the "new releases" aisle only to find it had been decimated and nothing left? "Want to see *Jaws* again, honey?"

I was always a bit anxious driving to Blockbuster, hoping there would be something new to watch that was in stock! At its peak, there were over 9,000 stores nationwide, and their goal was to have each strategically located to minimize your drive time and make themselves convenient. But they also had punitive late fees and, before DVDs, had fines for not rewinding VHS tapes.

...along comes Netflix.

Their model was to have strategically located distribution centers mail DVDs to customers and keep a large assortment of titles so that customers would almost always get their desired film or TV series. They mailed you what you ordered, and all you had to do was go to your mailbox to secure your entertainment. It would take two days, so long as you could plan. They include a return, postage prepaid envelope to return your DVDs, and no car drive is necessary; drop them in your mailbox.

They also changed their pricing model. Rather than charging per DVD, they deployed a subscription model enabling customers to order an unlimited supply for a single monthly rate.

They saved customers time and effort by eliminating the need to drive to the closest store and hoping to find a show they wanted to watch. They made not one trip but two using the United States Postal Service. Their subscription pricing model was cheaper once you ordered two to three monthly shows. The experience was great because you got to watch what you wanted and not leave it to chance that Blockbuster had your desired show in inventory.

I'm not sure how Netflix enabled buyers to trust them, but it didn't take long for Netflix to go viral and for Blockbuster to lose market share and close stores dramatically. Here's what is very interesting. After ten years, Netflix re-invented itself in 2007 by offering streaming services. Now, being more ***convenient*** than ever, a Netflix customer can instantaneously stream content. No more waiting a day or two for a DVD to arrive and no longer need to package the DVD and place it in the mail. It made itself even more ***economical*** because streaming allows customers to watch more content per month than using DVDs.

Had Netflix not initiated streaming services, I can assure you Amazon, which also offers streaming, would have done this and perhaps put Netflix out of business. Now, there are multiple streaming services, and a consolidation is happening.

The other interesting thing is that to enhance ***product experience***, Netflix began creating its own proprietary content, something no one else could copy. And because Netflix had been established with version 1.0, buyers could ***trust*** them to meet their expectations. Later, we will share the culture required to innovate continuously and meet buyers' needs to foster organic growth.

Several companies offer all four of the buyer's criteria we talk about. I like to refer to them as dominant companies because, compared to their competitors, they don't have a blatant trade-off or offer an excuse for a buyer to purchase elsewhere. They dominate the market share in their segment. Quik Trip is a regional convenience store company in that category, as is Nebraska Furniture Mart.

Quik Trip, the popular convenience store chain, follows a unique strategy to minimize buyer excuses and ensure customer satisfaction. When you enter a Quik Trip store, you will be impressed by its modern and clean appearance. The store offers various snacks, beverages, and warm customizable sandwiches, making it a one-stop shop for all your needs.

The staff, dressed in bright uniforms, is always ready to assist you with your purchases. The store is well organized, making it easy to find what you need quickly. The parking area is spacious, and the store is well lit, creating a safe and welcoming environment for customers.

Quik Trip takes great pride in maintaining the highest cleanliness and safety standards. The staff ensures fuel spills are promptly cleaned so you can shop without worries. They also stock up on windshield wiper fluid and paper towels so you can clean your car's windshield before hitting the road. Quik Trip's prices are always competitive and, in some cases, are aligned with grocery stores to incentivize grocery purchasing with discounted gasoline. They even offer a guarantee on the quality of their gasoline. This guarantee is a unique feature that other stores do not provide. Overall, Quik Trip is a convenience store that truly lives up to its name by providing quick and efficient service to all its customers.

So, they meet the criteria of being convenient, competitively priced, a great experience for a gas/convenience store, and exhibit trust with their gasoline guarantee.

2003 marked the opening of the Nebraska Furniture Mart, a massive retail and distribution space spanning over 1,000,000 square feet. This store offers an array of items, including furniture, home décor, appliances, office and outdoor furniture, carpets, and even lawnmowers. The store is owned by Berkshire Hathaway, a company known for its commitment to quality and excellence.

Despite having only one location, the store is conveniently located near two major interstates, making it easily accessible to customers. With ample parking space, the shopping experience is hassle-free.

Once you enter the store, you will be greeted by a well-dressed and polite associate who will guide you to the appropriate department and help you find your needs. The associates are equipped with tablets and can process transactions instantly without requiring you to wait in line at checkout. The store also offers warranties and affordable, if not free, home deliveries. If you're replacing an old item, they will even pick it up.

Nebraska Furniture Mart prides itself on offering competitive pricing, extensive selection, and excellent customer experience. Thanks to their ownership and warranty programs, customers can trust that they are getting a quality product at a fair price. Unsurprisingly, many furniture stores shut down shortly after opening as they could not compete with Nebraska Furniture Mart's offerings.

As a seller, being dominant is hard. Not many companies are dominant. Most have some trade-off where they focus on one or two of the buyer's criteria and sacrifice others. When you hit on all four, you, as a seller, are giving your buyers a few excuses not to buy from you. If you want unstoppable growth, strive to get there, and don't give up. I believe it takes a culture and environment where you are constantly looking to be more convenient, more price competitive, and transparent, offering a better product experience and being trustworthy for your product and the buying experience. You will also find that when you take away buyer excuses, you build loyal buyers who will tell others. Viral marketing is the best marketing you can have and minimizes the need for huge budgets to spread the word.

Summary

As we wrap up Chapter 8, we're left with a deeper understanding of what makes a company dominant in its field. It's not just about offering a product or service; it's about aligning every aspect of your business with the key values of convenience, price, product experience, and trust. This chapter has shown us that by focusing on these core areas, your business can stand out, attract loyal customers, and drive organic growth. Remember, the journey to becoming a dominant player in your market starts with a commitment to understanding and meeting your customers' needs at every turn.

Dominant companies excel in fulfilling all four criteria that buyers demand. They not only offer a wide range of products and services, but they also provide an exceptional experience for their customers. They understand the importance of minimizing the reasons for buyers to shop elsewhere and strive to provide convenience and simplicity in the buying process while offering transparent and competitive pricing. They build customer trust by providing guarantees, warranties, easy returns, and sometimes free or discounted trials.

It is a challenging task to fulfill all four criteria simultaneously. Many sellers prioritize revenue and profit over customer satisfaction and need to pay more attention to the importance of fulfilling the criteria.

In my experience, meetings often start with discussions about revenue, expenses, and budget rather than focusing on customer satisfaction.

If every seller prioritized fulfilling all four criteria, they could build long-lasting customer relationships and achieve tremendous success and growth, unstoppable growth.

CHAPTER 9

Trade-Off Companies

This chapter explores how companies balance convenience, price, product experience, and trust. Through engaging examples, we will uncover how different organizations position themselves based on these values, what this means for their competitive edge, and the risks they take. Business leaders and owners must understand the intentional trade-offs they present to buyers. Having any trade-off, be it a low price in exchange for an undesirable product experience or a high price for being very convenient, will limit a seller's ability to attract every buyer. Limiting the number of buyers, in turn, limits growth opportunities.

What is a trade-off company? Every buyer has their own individual personal value system, whereby they place specific buying criteria higher than others. Companies have their corporate value system that prioritizes the specific value or criteria on which they place the most emphasis.

For some buyers, we place the most weight on convenience when deciding what and from whom to buy. Those who value time, effort, simplicity, and ease will decide mostly on those convenience factors. Others place a higher priority on price when making a buying decision. Their value system tells them to buy the best deal or lowest price and prioritize that over the other factors. Then, other buyers will make decisions mostly based on product experience and weight more than price or convenience. Finally, there are buyers whose value system dictates they buy based on trustworthiness over everything else.

Let's move to the seller's perspective. Some companies focus on a single proposition, largely ignoring or lowering the value on other buying criteria. Other sellers try to fulfill more than one value or buyer criteria, but for now, let's keep it simple and assume one. When they focus on delivering one buyer value or criteria, they likely are trading off that one for other values or criteria.

How about a few examples? Convenience stores, by their name, cater to buyers who value convenience, saving them time and effort. They aren't worried about being priced competitively, nor do they necessarily go to great lengths to provide an outstanding product experience or promote their trustworthiness. They typically are located where there's easy access, parking close to the door, and a smaller store footprint supporting the ability to come and go quickly. They have limited selections and focus on items that buyers want who are on the move.

It's common for convenience stores to sell gasoline, coffee, snacks, beer, cigarettes, lottery tickets, and other relatively low-value items. They like to be located just off busy highways or near busy intersections, attracting travelers who are in a hurry but might need gas or want something to drink or snack on. They save buyers time, and effort and are simple and easy to access and make a purchase.

The quick oil change and auto lube industry also qualifies as a trade-off industry. They cater to and seek vehicle owners who value their time and don't want to exert the effort to change their oil. Jiffy Lube makes getting service without an appointment simple and easy, and they can complete their service in under 30 minutes. They are priced higher than options like Walmart or other auto maintenance businesses because they are trading off the value of convenience for price. The product experience is okay, but that isn't what they promote, and they don't actively promote anything that makes them more trustworthy than any other option, other than maybe a well-known brand.

Now, let's examine companies whose primary value proposition is their low price. They promote their pricing as their fundamental value proposition and may or may not offer convenience, a great product experience, or promote trustworthiness.

Stores containing the word "dollar" in them, such as Dollar General, Dollar Tree, and Family Dollar, come to mind. Via their name, they indicate that everything they offer sells for a dollar, but that is only in perception. Even though their primary value proposition is based on low-priced merchandise, they offer some convenience. You might find any of these stores in more rural locations, especially towns and cities that aren't large enough to support a Walmart or even a full-size grocery store. I think that is one of their strategies to locate in less populous areas and serve those who typically must drive some distance. I know that Dollar General offers limited grocery items, making purchasing critical items more convenient without having to drive into larger towns. The footprint of a store is about the same size or slightly larger than a convenience store. Still, they sell general merchandise, consumer goods, and limited grocery items.

The significant trade-off for these dollar stores is the product experience. They sell on price first and convenience second, and the product experience is less than ideal. They are often cluttered as they don't have back rooms to store incoming orders, they aren't staffed but with one or two employees, have just a couple of check-out lines that are not always staffed, and are not always spotlessly clean. They have a frequent staff turnover, and the stores aren't consistent with their layout and are not well-marked when searching for product categories. They are trading off price and convenience for product experience.

Aldi is a well-known grocery store chain offering extremely low-priced, mostly private-label groceries and odd merchandise. They cater to the buyer who values low prices over the other criteria buyers have. Their store footprint is much smaller than conventional grocery stores, and they are primarily located in larger towns and cities.

While the price point is their strong suit and moderately convenient, the product experience is an absolute turn-off for those who place product experience as their top value or criteria. You must bring your shopping bags or buy them from Aldi when you shop at Aldi. If you want to use a shopping cart, bring a quarter to unlock a cart and return it to collect your twenty-five cents.

Aldi has a tiny produce section, no deli, no butcher department, and very few national brands are sold there. They are a prominent private label seller, and without all the frills of free shopping bags, less labor for collecting carts in the parking lot, and specialty departments, the product experience is their primary trade-off.

Allegiant Airlines is a discount airline focused on travelers who value price as their primary buying criterion. Their direct trade-off is product experience, but convenience is an almost equal trade-off. Allegiant often uses secondary airports to service, often within significant metro areas. For example, they service Tampa/St. Petersburg, but they only fly into St. Pete, where virtually no other airline flies. They utilize the Sanford, FL airport near Orlando and not Orlando International. They only fly on limited days to service origin-destination pairs, such as servicing St. Petersburg to Kansas City, maybe on Tuesdays and Thursdays only.

Allegiant uses a somewhat deceptive advertising scheme, so while they appear low cost, one must be careful as you must pay extra for items such as any seat assignment, onboard beverages, carry-on items, checked bags, and priority boarding. I've seen fares advertised as low as $37 one-way, but by the time you arrange a seat, carry on an item, and buy a beverage, the actual cost will be closer to $150. While they are a safe airline, the seats are not as cushioned nor have as much legroom throughout as other airlines. But, for travelers who place low prices above convenience and product experience, they are a good option.

Walmart is a very well-known and popular retailer that is an operationally efficient company that promotes a value proposition of "everyday low prices," "helping people worldwide save money and live better," and the famous "Rollback America" slogans. Walmart's primary trade-off is product experience. They are becoming more convenient with curbside pickup and home delivery programs. However, while Walmart offers an extensive assortment of products and groceries in very large stores, the one experience that perturbs users is their checkout experience.

Even before the self-checkout phenomenon, Walmart was known for having only 20-25 checkout lanes staffed with maybe three cashiers, forcing people to wait in long lines to pay. With self-checkout, there is significant frustration because many shoppers believe that a cashier checking them out should be part of the service and paid for with the margins of the goods. It's incredibly frustrating using self-checkout for certain produce items. I tried to buy apples and use the self-checkout and found out there are three pages of different types of apples. I didn't remember what apple I selected. Then, if buying alcoholic beverages, you must wait for an attendant to validate the buyer's age, and attendants aren't necessarily very accessible. And now, they are using AI cameras to ensure you aren't skipping an item when scanning or placing a sticker from an inexpensive item on an expensive item. And if you make a mistake, such as scanning an item twice, your checkout experience is going to cost you a lot of time. Recently, like Costco, they now check your printed receipt to ensure you didn't steal or forget to scan something. It's not a great product experience.

Given the size of Walmart stores, parking lot size, and store layout, it is not incredibly convenient, but the company is trying to make up for that with its curbside delivery and online shopping features. Walmart tries to strategically locate stores close to population density, but although a drive to Walmart might be quick, shopping and checking out are not. They target those who prioritize price and trade-off product experience and convenience.

Nordstrom is a retailer that caters to the buyer and places product experience at the top of their list. Their stores are well-appointed and well-lit, their staff is courteous, and their merchandise is often top-of-the-line. Nordstrom can even arrange personal shoppers to help those who like that touch. Their primary trade-off is their pricing. When buying from Nordstrom, you will spend more money than with other options. Nordstrom is also not necessarily known for their convenience. They are in larger metroplexes and sometimes target malls to set up shop, which is inconvenient.

Cadillac is an automotive brand seeking a buyer who places product experience at the top of their value system. A Cadillac will be stylish, comfortable, large, and spacious and offer as smooth of a ride as possible. The auto company is not pretending to be the choice for those seeking price leadership. They also aren't catering to those seeking convenience. When you opt for Cadillac, it's all about style, comfort, and status.

Four Seasons is a high-end hotel and resort chain that offers many amenities, including large rooms, comfortable beds, stylish décor, soft and cushy towels, and restaurants and bars. They target those travelers seeking an exceptional product experience, whether on vacation or staying the night. Their trade-off is price and convenience. Your final cost at any Four Seasons is not anywhere near a Motel 6. They also aren't catering to those looking for convenience. You will rarely, if ever, see the Four Seasons along an interstate highway seeking the overnight car traveler.

All these examples are meant to point out that many companies and sellers consciously seek specific buyers that match their top value proposition to their personal value system. They are consciously not trying to cater to all but trade off one value for another. That does not make them bad companies or poor sellers. Those that I highlighted are great companies and do what they do well. The key is that they don't attract buyers whose value system makes another criterion more important. They target, advertise, and market to a specific segment of buyers, and that can limit their growth and give certain buyers an excuse to buy elsewhere.

Buyers who place price competitiveness at the top of their value system will not be most attracted to convenience stores, Jiffy Lube, Nordstrom, Cadillac, or the Four Seasons. Those who place convenience at the top of their list aren't going to be immediately drawn to Aldi, Walmart, Allegiant Airlines, Nordstrom, Cadillac, or Four Seasons. Those who place product experience at the top of their value system aren't seeking out convenience stores, such as Walmart, Aldi, and Allegiant, to shop.

When I evaluate most sellers or companies, it's easy to identify the tradeoff. Most argue they can't be all things to all buyers. If they are going to offer convenience, then the buyer must pay for that. Or, if the seller is going to provide the very best product experience, the buyer will pay for that, etc.

Dominant companies deliver on all four of the buyer's values and criteria. They offer convenience, are priced transparently and competitively, provide an exceptional product experience, and are trustworthy. In other words, they minimize and work to eliminate the excuses buyers have for not doing business with them.

Here are examples of excuses one might hear:

> *"I would buy my lottery tickets from the convenience store, but the cashier is behind bulletproof glass, and it's not well-lit and clean."*
>
> *"I would buy a Cadillac, but they are too expensive."*
>
> *I would buy from Walmart, but it takes too long to check myself out or wait in the cashier line."*
>
> *"Why would I stay at the Four Seasons when I can stay at a Holiday Inn Express?"*

I was talking to the CEO of a sheet metal fabricator. He explained that their company couldn't be the low-cost provider because their shop was unionized, and their fully loaded labor cost, compared to non-union competitors, was at a 30% disadvantage.

While you limit your audience of buyers, you don't have to be the low-cost provider to win at business and grow. The key is being exceptionally strong at delivering to the other buyer values. If low cost or price won every time, there would be lines around the block to get into every Aldi's. The Four Seasons would have been put out of business by Motel 6. No one would be driving a Cadillac but used Kias with over 100,000 miles. No one would fly first class.

Taking this into account, I suggested that CEO save their customers time and effort in the sheet metal fabrication case. Deliver bids faster than competitors. Promise delivery faster. Make ordering and engaging easy. Make sheet metal fabrication more convenient than that of non-union competitors. From a product experience perspective, collaborate and suggest ways to help the buyer succeed. Call them and keep them updated on every order. Ask if there are other ways to help them save time and effort. Keep them informed of every milestone for each job. Guarantee the product. Provide a better warranty. Guarantee their satisfaction!

In other words, provide as many of a buyer's criteria and value as possible. Do not rest on a single value proposition. The more buyer criteria you can satisfy, the fewer excuses a buyer will have to buy elsewhere.

Here are two suggestions I make to any seller. Many times, sellers will go to buyers and ask how they can win more business from the buyer. "What can I do to win more business?" I wouldn't say I like that question. Instead, I would ask buyers: "How can we save you time?" "How can we save you effort?" "How can we make buying and using us easier and simpler?" The little things add up, and by asking these questions, you are telling the buyer you care about them and want to save them time and effort and make their lives simpler and easier. And, when you deliver, they appreciate it and can justify paying more.

Any seller has the most control over the product experience they deliver. Buyers want to be treated like they are the most critical customer a seller has. How difficult is it to treat them that way? How hard is it to be proactive with them, check in with them, and ask how you are doing for them? Tell them how important they are. How hard is it to have your employees look professional and dressed in uniform, if appropriate? How hard is keeping your building organized, clean, and well-lit? Think about what you like in an experience and do it!

It's extremely hard to offer all four of the buyer's criteria. Most companies and sellers have some trade-offs, and as stated, that is fine. However, if you are a seller and hanging your hat on only one of the criteria, you risk a competitor knocking you off. You also limit your growth as you are catering and attracting only those buyers who match their personal value system to your primary value proposition. If you want to grow, examine how you rate on all four and do everything possible to offer as many of the buyer's criteria as possible. It may not be easy or happen quickly, but you should do everything possible to give the buyers what they want.

As we conclude this chapter, we have gained a deeper understanding of how successful businesses strategically prioritize different buyer values. This chapter challenges us to think critically about our business strategies and how we can effectively balance convenience, price, product experience, and trust to meet our customers' needs. Remember, in the dynamic business world, the ability to adeptly navigate these trade-offs can be the difference between merely surviving and thriving. As you move forward, let the insights from this chapter guide your decisions, helping you to build a business that not only meets but exceeds the diverse expectations of your customers, fueling sustainable organic growth.

Summary

Buyers, whether individuals like you, me, or professional corporate buyers, all have value systems. Buyers have four values or criteria by which they make purchasing decisions; everyone is different.

Trade-off companies intentionally pick a value and promote and leverage that in place of the other values that buyers have. Convenience stores aren't price-sensitive and don't care so much about the product experience. They are driven to deliver to the buyer who places convenience at the top of their product list. Jiffy Lube has a trade-off that emphasizes convenience over price.

Walmart, Aldi, and Allegiant cater to and serve buyers who value price over convenience and product experience. Nordstrom, Cadillac, and The Four Seasons are positioning product experience ahead of price and convenience.

If possible, sellers must cater to and satisfy as many of a buyer's criteria and values as possible. For example, if you cannot compete on price with someone like Walmart, be certain to be convenient, provide an exceptional product experience, and be more trustworthy. The goal of any seller should be to minimize and eliminate the excuses buyers use to buy from you.

The easiest buyer criteria or value to fulfill is product experience. Sellers are more in control of providing a great product experience than they are with convenience, price, or trust. There should be no excuses for not providing your buyers with a great product experience.

CHAPTER 10

A Lesson from Cheese

In this chapter, we explore how businesses, like grocery stores, must accommodate their customers' diverse needs, balancing convenience, price, product experience, and trust. From the dairy aisle to the digital revolution, we will uncover the secrets behind successful businesses like Netflix and Uber and how their ability to cater to these four critical criteria has fueled their exponential growth. Get inspired by these stories and apply these lessons to achieve unstoppable organic growth in your business.

Do you ever buy cheese? If you are a price-conscious shopper whose personal value system drives you to buy based on price, then there is a preferred cheese option. The most economical form of cheese is a block of cheese. No matter what flavor or size you prefer, buying a block of cheese is the most economical option on a per-ounce basis. Plus, a block of cheese is a versatile ingredient that can be used in many ways.

You can slice it to create the perfect sandwich or snack with crackers. Cubing it will make it easy to pick up and eat as an appetizer or snack. Shredding it will allow you to use it in your favorite recipes or sprinkle it on top of your favorite dishes, tacos, or salads. You can cut it into cheese sticks and put them in your lunchbox for a delicious snack.

Walk to any dairy section in a grocery store, and you will see many convenient options: different brands, flavors, varieties, and styles.

In addition to blocks, sliced cheese options come in different sizes, perfect for sandwiches or crackers. Individually wrapped slices make grabbing just one slice at a time effortless. Cubed cheese is pre-cut into bite-sized pieces, ready to be added to your favorite recipes. Shredded cheese comes in various blends, perfect for cooking or topping your favorite dishes. Cheese sticks are individually sealed for maximum freshness and are ideal for snacking.

If you want to enjoy a delicious cheese platter, you have a few options. You can buy pre-sliced and pre-shredded cheese for added convenience, purchase a block of cheese, cut dice, or shred it yourself. Pre-cut cheese options will cost more per ounce than a block of cheese. However, if you're on a budget, it may be worth the extra time and effort to cut, dice, or shred your cheese.

While cutting your cheese may take more time and effort, it has benefits. You can cut your cheese into the exact shapes and sizes you want, although you may need help to achieve the same thin slices you'd get from store-bought cheese. You can use a traditional hand-held shredder or an appliance to shred your cheese. Cubing cheese is also an option, but it will require more time and effort. If you don't care about having equally sized cubes, it's a good choice.

If you want to preserve your cheese sticks, you can use sealable sandwich bags. While this method may not be as effective as a manufacturer's vacuum-sealed solution, it can help maintain the freshness of your cheese. However, it's important to note that there will be some clean-up involved, as your knife, grater, or appliance must be cleaned thoroughly. Also, depending on the type of cheese you're working with, you may experience some crumbling or loss. So, be prepared to put in some effort if you cut, dice, or shred your cheese. Those who buy blocks will likely buy them for the price or economic benefits. Those who buy pre-prepared cheese mostly buy it for convenience and product experience.

When buying cheese, most buyers tend to opt for the already-prepared cheese based on their intended use. They are willing to pay a bit more per ounce for the convenience of not having to slice, dice, or shred the cheese themselves. However, it's always good to have options available.

You may be wondering about my point and why the discourse on cheese.

Here's the point: I have many options as a buyer in a grocery store's dairy section. I have many options if my personal value system is driven by convenience. If my unique value system is driven by price, I can buy the block of cheese that costs the least on a per-ounce basis. If I am looking for product experience, the cheese that allows me to seal or only use individually wrapped pieces allows that. I might opt for the national brand to buy cheese I trust.

As a seller, offering your buyers ***multiple product options*** is crucial to attracting a wide variety of buyers and satisfying each buyer's personal value systems. The more options you provide, the fewer excuses the buyer will have not to buy from you. Various options mean considering the extra costs you may incur in providing a product that makes it easier for buyers, saves them time and effort, and offers a better product experience.

There are different ways you can go about this. If you offer a premium product at a premium price, consider adding a product line that costs less but requires the buyer to put in more work. An example is the meal kit industry, which provides customers with high-quality raw materials, exact portions, and detailed instructions to cook a restaurant-quality meal at home. It's as good as eating at a restaurant while saving you time and effort by not traveling to the restaurant. Meal kits are an option and ideal for those seeking the quality of eating at a restaurant but more convenient.

You can always find recipes and purchase all the necessary ingredients when cooking from scratch. You may need to buy spices or condiments you don't use for more complex meals, leaving them on your shelf or in your refrigerator for an extended period. The bottom line is that meal kits offer an option for a pleasant experience at a lower cost but require time and effort. The meal kit industry has filled the void between going to a restaurant, buying groceries, and cooking, offering people the chance to enjoy a high-quality meal at home without the hassle.

The larger grocery stores often offer cooked items, like fried chicken or meatloaf. They have pre-arranged cheese and deli trays. They might have salad bars or small grills. It's a way to leverage their core competency but offer options that fit multiple buyers' personal value systems.

Airport parking has learned the lesson of cheese. I'll take my home airport of Kansas City as an example. Their block of cheese is their discount parking, located a few miles from the terminal. It has non-covered parking and roughly a dozen stations where a bus regularly stops, picks up, and drops off passengers. Drivers find an empty parking space, gather their suitcases, walk to the nearest station, and wait for the next bus. Depending on your location, you must make several stops until the bus arrives at the airport. On the other hand, when you arrive at the airport, you go to the bus pickup station and wait for the bus to take you back to your station, and you walk to your vehicle.

If it's cold, hot, or raining, you must endure the elements and walk a bit to get to the airport's lowest car parking cost. If you prefer not to ride the bus, they have parking within walking distance of the terminal, but it's non-covered and not as close as the parking garage. Then, if you want to spend more, you can park in the airport's parking garage and walk along an enclosed walkway to the terminal. Finally, the best product experience is the valet parking option, where you drop off and pick up your vehicle at the curbside. It's nearly 50% more than the garage parking.

Like cheese, the airport provides multiple options with various convenience, price points, and product experiences. There are different trust levels, too, as you need to know precisely how much time the remote lot will take to get to the terminal, how many spaces are available, etc.

Summary

When selling products, offering a low-cost option may only satisfy some buyers. It's crucial to develop a range of options that provide a better experience, save time and effort, are simple and easy to use, or offer more of a guarantee. The more options you offer buyers, the more chance they will not find an excuse to go elsewhere.

While some buyers may prioritize saving money over quality or convenience, others may be willing to pay more for a superior product experience. Therefore, it's essential to offer various options that appeal to both types of buyers. These could mean providing a low-cost option for budget-conscious customers but also offering a more expensive option that provides a seamless product experience, saves time and effort, and comes with a guarantee.

To stay competitive and to appeal to wider buying audiences, it's essential to constantly seek ways to improve your products by making them more convenient, lowering costs, making pricing more transparent, enhancing the product experience, and assuring buyers that their expectations will be met. After making changes to improve convenience or other aspects, it's imperative to keep evaluating how to be even more convenient and better meet customers' needs. Failure to do so may result in your competitors overtaking you, and your growth may not be sustainable. Starting a business or expanding an existing one can be daunting. Identifying opportunities that resonate with customers, providing value, and differentiating your offering from competitors is essential.

CHAPTER 11

Launching and Expanding Products

In this chapter, we will examine the logic and strategies often used to either start a business, or to expand product lines.

As an aspiring entrepreneur or a company looking to expand, it's crucial to look beyond your skills and passions and focus on customers' needs. Instead of starting a restaurant because you enjoy cooking, discover gaps in the market and opportunities to offer products or services that solve customers' problems. By identifying businesses or industries that underserve their customers, you can create a unique value proposition and gain a competitive advantage.

To identify underserved markets, you must look for products or buying processes that are too time-consuming, complex, overpriced, or not transparent in pricing. These businesses offer inferior product experiences without guarantees, warranties, or easy returns. By providing a better product experience, transparent pricing, and customer-friendly policies, you can create a competitive advantage and gain customer loyalty.

If you're already in business, revisiting your value proposition and ensuring you're meeting your customers' needs is essential. It would be best if you focused on delivering on as many buyer criteria as possible, including convenience, price, experience, and trust. Look for opportunities to expand and offer new products or services that appeal to a broader band of buyers. By listening to your customers and understanding their needs, you can develop innovative solutions and gain a competitive advantage in the market.

Too many people start businesses because it's either their passion or they have thorough industry knowledge and a self-belief that this knowledge will be the driving force instead of analyzing and understanding buyer needs.

I did it once. I started a trucking company simply because I knew the trucking industry. I didn't understand where there were shortcomings by existing trucking companies and how I could entice customers to buy from me because I offered them a more convenient service, better and more transparent pricing, a unique product experience or trust. Guess what? My startup failed relatively quickly.

This lesson is intended to benefit both established companies and aspiring entrepreneurs. As mentioned, starting a company solely based on expertise is required. For example, opening a restaurant because of a passion for cooking or starting a trucking company because of a love for driving trucks is insufficient.

Searching for businesses or entire industries that fail to deliver a satisfactory product experience is prudent for budding entrepreneurs. These businesses may offer a product or buying process that is too time-consuming, requires too much effort, is overly complex, overpriced, or needs more transparency in pricing. Additionally, consider products that lack guarantees, warranties, or easy returns. By identifying these gaps, you can make an impact and gain traction with your product. Look around at new businesses starting in your community. Did they start a business because of a gap in meeting the four buyer's criteria?

Let's examine a few recognizable brands that did it the correct way, being having a product that filled the needs of buyers. Both Netflix and Uber launched with one business idea in mind but progressed and continued to figure out how to be more convenient, lower prices, improve product experience, and gain trust.

Netflix

Netflix was founded in 1997 by Reed Hastings and Marc Randolph in Scotts Valley, California. The origin story of Netflix is exciting and has become legendary. It's said that Reed Hastings came up with the idea after being charged a $40 late fee for a VHS rental of "Apollo 13" from Blockbuster. Whether that's entirely true or a bit of entrepreneurial mythmaking, the frustration with late fees was a real issue for many, and it certainly helped in shaping Netflix's initial business model.

The company started as a DVD rental service, where customers could order DVDs online and have them mailed to their homes, a novel idea at the time. This model was revolutionary because it eliminated the need for physical rental stores, taking the time and effort to drive to a store and then return the DVD and face potential late fees, all of which were significant pain points for buyers. The negative was the pre-planning it took to order a movie and allow the Postal Service the time it took to deliver.

Netflix also modified its pricing strategy from a per-movie or DVD rental to a subscription-based model, providing an economic incentive for those who watched multiple DVDs monthly.

However, they streamlined by making their offering more convenient and a better product experience. What set Netflix apart was its adoption of new technology and its willingness to pivot its business model. In 2007, Netflix began offering streaming services, allowing users to watch TV shows and movies over the internet instantaneously. They eliminated the inconvenience of retrieving and returning DVDs to the mailbox. This move was visionary, considering that high-speed internet was becoming more widespread.

Netflix's transition from DVD rentals to streaming was a significant risk, but it paid off tremendously. The company invested heavily in its streaming technology and content library. They also started producing original content, which began with the series "House of Cards" in 2013. This move into original content was another game-changer, making Netflix a distributor of content and a respected producer. Creating its content provided a unique product experience no one else could duplicate.

Today, Netflix is one of the leading streaming services globally, known for its vast library of films, documentaries, and television series, including many critically acclaimed original productions. Imagine if Netflix had initially tried to launch brick-and-mortar stores to compete with Blockbuster. Where would the differentiation be? Imagine had they not adopted streaming. Amazon could have put Netflix out of business or made them a second-tier offering.

By offering multiple plans, Netflix gives buyers options based on price level, product experience, and convenience. For buyers whose personal value system motivates them to buy based on price, Netflix offers a Standard Service with ads, making it their lowest-priced subscription service. They also offer Standard and Premium services, adding features to each. They understand the story of cheese and give buyers options, so they have fewer excuses to say no.

Uber

Uber was founded in 2009 by Travis Kalanick and Garrett Camp. The idea for Uber came about one snowy evening in Paris in 2008 when Kalanick and Camp, unable to get a cab, brainstormed a simple idea: tapping a button on your phone to get a ride. This simple idea was the seed that grew into Uber.

In 2009, they launched the service in San Francisco. The choice of location was strategic—San Francisco was known for its tech-savvy population and notoriously inadequate taxi service. The initial service, Uber Cab, offered rides in black luxury cars at about 1.5 times the price of a taxi. This premium service targeted a specific market segment willing to pay more for a reliable and upscale transportation option.

What set Uber apart from traditional taxi services was its user-friendly app, which allowed users to track their rides, receive fare estimates, pay through the app, and rate their drivers. This level of convenience and transparency was revolutionary in the personal transportation industry.

Uber's growth was rapid and aggressive. They employed a combination of technology innovation, fierce marketing, and sometimes controversial tactics to expand. The company was open to entering new markets, often flouting local transportation laws and regulations, leading to legal battles in various cities and countries. This aggressive expansion strategy was a double-edged sword, fueling their rapid growth and considerable controversy.

A significant part of Uber's strategy was using the gig economy model, where drivers were independent contractors rather than employees. This model allowed Uber to scale up quickly without the overhead costs associated with a traditional workforce and keeping fares low.

Today, Uber is more than just a ride-hailing service. It has expanded into food delivery with Uber Eats and freight transportation with Uber Freight. These expansions showcase its ability to innovate and diversify its business model in response to market opportunities and challenges.

Uber is more convenient, enabling riders to save time and effort versus conventional cab rides, and it is simpler and easier to engage than cabs. Its pricing is lower than cabs' and 100% transparent before taking the ride. It is a better product experience, riding in a car much like we all own, and can be trusted by the ratings other riders grade their drivers and service. It hits all the marks.

It started as a black car option but was adapted to be viable for the masses. Uber now offers various tiers of rides, from X, which is considered their basic ride, to XL, providing a bit more comfort but at a slightly higher rate, UberX Share, their more economical service and Uber Comfort featuring larger more luxurious vehicles. Like cheese, Uber has an offering to cater to nearly everyone's personal value system.

If you are already a seller and operating successfully, it's essential to ensure that you meet as many buyer criteria as possible. Then, focus on expanding your offerings to appeal to a broader range of buyers. Buyers have different value systems, and their priorities may vary based on convenience, price, experience, and trust. Identify the gaps in your product offerings based on the feedback from your buyers. Capture and track their feedback to make informed decisions.

Often, companies need to adequately consider the buyer's needs to expand their product offerings. Instead, they focus on their competencies. However, to meet buyer demand, it's crucial to understand what drives buyer behavior and leverage your competencies accordingly. You can offer a product that meets your customers' needs while maintaining your competitive edge.

As we conclude this chapter, the lesson from cheese becomes a profound insight into business strategy and customer satisfaction. This chapter has shown us the importance of understanding and catering to various buyer preferences and highlighted the dynamic nature of the business landscape, where adaptation and innovation are key. The stories of Netflix, Uber, and others remind us that success comes from continuously evolving to meet and exceed customer expectations.

As you continue your journey in the business world, remember that, like cheese, your offerings must be versatile and appealing to a spectrum of tastes and needs. This adaptability and focus on customer values will propel your business towards organic growth and long-lasting success.

Later in this book, we will delve into the necessary steps required to nurture a culture that facilitates the successful development of a product that meets the four vital criteria of a buyer. It's an all-hands-on-deck approach where everyone in the organization must be involved to achieve success. Without collective effort, even the most visionary CEO's plans may falter.

Summary

Whether starting a new venture or expanding the product mix of an existing enterprise requires the seller to understand the shortcomings of those already in the market. It's about more than having a knowledge or passion of a certain industry.

Netflix is an excellent example of a company that successfully transformed the market by introducing streaming and producing original content. This move made their product more convenient than the conventional DVD mailing system, offering a product experience that surpassed all others. Had Netflix not enhanced its product, a competitor like Amazon would likely have overtaken them, and they would have ended up like Blockbuster.

Uber is another example of a seller that created a more convenient, less costly service with a better product experience that everyone could trust. Netflix and Uber started by catering to buyers' needs and their four criteria. The most important lesson is that neither rested after the initial business model design. They improved upon their original model to better meet buyer needs.

Entrepreneurs and sellers should always look for opportunities to develop new products or companies based on emerging trends and weaknesses in the market. By doing so, they can leverage their strengths and stay ahead of the competition. Giving a buyer more options with various buyer's criteria being fulfilled is a way to eliminate or minimize the excuses buyers have for not purchasing from you.

CHAPTER 12

Customer-centric Culture

Now, let's explore the pivotal role of placing the customer at the heart of every decision and action. We'll look at leading companies like Amazon and Apple, dissecting how their relentless focus on customer needs has propelled them to extraordinary heights. But this isn't just a story of the giants; it's a blueprint for businesses of all sizes, from startups to established enterprises, to cultivate a culture prioritizing the customer above all else.

We have discussed the strategies to be implemented for growth, but without the proper supporting culture, a company may be unable to sustain consistent expansion. Companies can hire consultants for projects, but because an outsider comes in and makes or recommends changes, those changes aren't always endorsed and supported over the long term. In this case, embracing a customer-centric culture will enable the entire organization to adopt the values and behaviors to sustain and improve the growth strategies required. The strategies seem simple, but I can assure you it will take consistent behavior and improvement over time to gain the most benefit.

For example, introducing new technology will likely impact operating efficiency, leading to lower operating costs, which can be leveraged for more competitive pricing. New technologies can also improve the convenience factor (that is, save time and effort) or make tasks simpler and easier for buyers. New technologies can also greatly improve the overall product experience.

If one doesn't have the culture to identify ways to improve the overall buying experience constantly, a company can be easily surpassed by those that do. Technology can have a purpose, be part of a culture, or be leveraged to increase efficiency, margins, and profits, but the ROI on technology investments are much more significant when the technology enables the buyers to meet their four buying criteria.

While I don't consider myself an authority on company or business culture, I have developed strong beliefs about creating a customer-centric culture. In my career, I've worked for companies solely focused on operational efficiency, which often resulted in a lack of focus on the customer experience. I am familiar with cultures that focus on product innovation but might fail to meet customer expectations in meeting buyer price points. The best culture to support growth is a genuinely customer-centric culture, where everything the company or seller does is targeted to meet all four of the buyer's buying criteria.

I am 100% convinced that most companies don't have an identifiable culture. They were founded to sell products, and that is what they do, but they don't have everyone on the team aligned to a set of values or missions other than to sell their products.

Most companies exist without a unique culture that sets them apart from others. While some companies may offer employee-focused perks such as free lunches, happy hours, beer-thirty Fridays, casual dress, dogs allowed, or work-from-home opportunities, these activities may only sometimes create an identifiable culture. Those practices are more geared toward appeasing employees and making their lives easier or perks to try and lure employees to work there, but they don't have much to do with culture. Instead, a company's culture is shaped by its core values, beliefs, and practices.

Customer-centric cultures are all about the value, belief, and practices of placing the buyer as the center of everything they do and engineering backward to ensure the buyers get what they want. At the same time, the company also gets what it wants.

Companies want growth both in revenue and profit. When they can grow their revenue and profit, everyone wins, including the buyers, as intelligent companies re-invest in the overall customer experience and fulfill the four criteria buyers use to make decisions.

I am not going to pretend that I am an expert on culture. However, I have worked for multiple companies and engaged with many more. Using the definition of business culture as the shared values, beliefs, and behaviors that characterize a business, most companies don't have a defined culture.

For example, if you ask any company employee to share their company's shared values, beliefs, and behaviors, you might get a lot of blank stares. It is similar to Simon Sinek's' book *Start With Why*©. Some have it, and some don't. Sinek is a disciple of having a why or a purpose, and everyone buys into that purpose. I agree. But all too often, if you poll workers at a company and ask them about their "why," you'll likely get a blank stare or an answer such as "to get a paycheck."

In *The Discipline of Market Leader*©, authors Michael Treacy and Fred Wiersema present a clear framework for businesses aiming to dominate their market. The book, first published in 1995, argues that companies must focus on one of three value disciplines to achieve market leadership: operational excellence, customer intimacy, or product leadership. Operational excellence involves streamlining operations and delivering goods and services more efficiently and consistently than competitors. Customer intimacy focuses on providing good products and tailoring and personalizing services to meet individual customers' unique needs and desires. Product leadership, conversely, centers on innovation, offering leading-edge products and services that push the boundaries and redefine the market.

The authors emphasize that while companies must maintain threshold standards in the other two disciplines, authentic market leadership requires excelling in one area. The book is filled with case studies from various industries, illustrating how successful companies have focused their strategies around one of these disciplines. Treacy and Wiersema's model has influenced business strategies across sectors, advocating for focused excellence rather than over-diversification. The key takeaway is that in a competitive and fast-changing business environment, the discipline of focusing on and mastering one key area of value creation is vital for companies looking to achieve and maintain market leadership.

If I were to categorize company cultures, I would emulate what Treacy and Wiersma preached in 1995. In other words, there are companies with a customer-centric culture focused on product innovation and a culture centered on operational excellence or efficiency. But, true customer-centered cultures focus on the entire customer experience, not just solely innovation, experience, or efficiency for the customer. The old way of thinking was to focus on a single area but in today's saturated business world, a seller looking to grow must push beyond only one and cater to *all* of a buyer's needs.

Amazon is an excellent example of a company that embraces a customer-centric culture. Apple is a company that constantly strives to innovate, improve upon existing products, and create new products. UPS and Walmart have a culture that emphasizes operational excellence and efficiency. Each is highly engineered to drive highly efficient results that are more evaluated on internal measurements than external customer metrics. Having a culture is better than not having one, but I can't say that one culture is better than another one.

Companies that excel in product leadership firmly believe that the byproduct of their culture is satisfying customers by delivering products that provide convenience and a better product experience. Companies focused on efficiency believe their product satisfies customers via an exceptional product experience and lower pricing. Rather than heavily engaging customers to learn, they make decisions internally on what they believe the customers want.

Take UPS, for example. Their focus on operational efficiency is evident in their highly engineered processes and tight performance standards. Their fundamental belief is that by being efficient, they can deliver packages on time and at as low a price as possible, which is surely what customers want from a delivery company.

Walmart is an operationally efficient company, and its internal belief is that offering various goods at low prices delivers what buyers want.

Starbucks, on the other hand, is highly focused on customer experience. The company goes above and beyond to provide a comfortable and welcoming atmosphere for its patrons, complete with a wide variety of great-tasting coffee flavors and free Wi-Fi to encourage customers to stay and enjoy their space.

As someone who has spent over 40 years working for a dozen or so different companies, I have encountered many that claim to be customer-centric. Unfortunately, in my experience, these claims often become nothing more than empty rhetoric. They surmise what their customers want and don't engage in formal processes to listen to them and understand their changing needs.

I can assure you that during my working career, I have not attended a meeting where the objective was to focus on how many new customers we gained and why. Or how many customers we lost and why. Rather, meetings focused on internal metrics such as revenue, costs, budgets, IT backlogs, hiring, and capital expenditures. Meetings were always focused internally on what we, the company, should be doing and not what the buyers wanted.

If a company truly prioritizes its customers, shouldn't it be discussing key performance indicators related to customers in its meetings?

Sure, many sellers measure and discuss Net Promoter Scores (NPS), but they don't deeply analyze why they win and lose customers and set up tactics to address each.

If a company is truly customer-centric, it would prioritize gaining and retaining customers above all else. This culture would mean regularly discussing customer KPIs and working to improve the customer experience. After all, customer-centric companies understand that they survive only because of the revenue buyers generate. Revenue, in turn, drives profit. Profit drives growth and everything the company does.

Amazon is a company that I cannot help but admire for its remarkable customer-centric approach. Their focus on customer satisfaction is their top priority. I remember applying for a job at Amazon a few years ago. My background in the logistics industry made me appreciate Amazon's logistics prowess, which is second to none. While they sell merchandise, their primary focus is on logistics—offering a vast array of products for sale and ensuring that they are delivered promptly to the customers.

As part of my preparation for the initial interview, I was sent Amazon's 16 Leadership Principles, which their leaders live by. The first principle, which is the most important, emphasizes the importance of customer obsession. It is evident that Amazon takes this principle seriously and strives to provide the best possible experience to its customers.

> *#1 – Customer Obsession.*
>
> *Leaders start with the customer and work backward.*
>
> *They work vigorously to earn and keep customer trust.*
>
> *Although leaders pay attention to customers, they obsess over customers.*

The term "obsess" holds a special place in my heart. It represents an unwavering commitment to constantly thinking about something.

It's a powerful word with more weight than "care" or "consider." Specifically, when it comes to the customer, obsessing over their needs and desires is paramount.

I truly appreciate the idea of starting with the customer and working backward. It means designing and engineering products and services with customer needs in mind. Unfortunately, many companies focus on their own capabilities and hope customers will find their products desirable. They rationalize that because they are good at producing something, there must be customers who will appreciate it and buy it.

Amazon's customer-centric philosophy is a shining example of how businesses should operate. They understand that if you prioritize the customer, other aspects like revenue and profit will naturally follow.

Moreover, putting the customer first is a crucial aspect of this philosophy. It means obsessing over the customer's needs and desires and working backward to fulfill them. This approach is much more effective than building a product and hoping customers will buy it.

Ultimately, the customer should always be our top priority. We must structure everything around taking care of and satisfying the customer. By doing so, we can build a successful business that creates value for both the customer and the company.

In 1984, I embarked on a new career path by joining the esteemed Yellow Freight System in Springfield, Missouri. As the night shift supervisor, my primary responsibility was to oversee the processing of incoming shipments that needed to be routed and delivered on a timely basis.

After two years in that role, my dedication and hard work paid off, and I was soon promoted to the position of Operations Planning Analyst at the company's headquarters in 1986. It was a tremendous opportunity for me to showcase my skills and gain valuable experience in the industry.

After five successful years in this role, Yellow announced plans to launch a new division called Yellow Logistics. The logistics concept was still relatively new in the trucking industry, and I was excited to learn more about it. Our goal was to provide customers with a comprehensive service enabling them to manage their entire shipping process by routing and managing their shipment activity.

Yellow founded Yellow Logistics to compete with its two biggest rivals, Roadway Express and Consolidated Freightways, which had already established logistics subsidiaries named Roadway Logistics and Menlo Services, respectively. However, no clear strategy or culture was in place. In my opinion, the only strategy they had was to mimic its competitors. That awful strategy ultimately led to the demise of Yellow Logistics, and it was shuttered. The parent company felt that the logistics division conflicted with the trucking organization. Today, billions and billions of revenues are generated by logistics companies as they cater to the specific needs of shippers, offering them convenience, a better product experience, and lower costs. Had Yellow been focused on what the customer or buyer wanted; they would have likely succeeded.

To be truly customer-centric, successful companies need to center their universe around their customers and ensure they meet their ever-evolving needs. This culture requires companies to continually modify their Key Performance Indicators (KPIs), listen attentively to customer feedback, and be agile enough to adapt to changes in the marketplace. The focus areas for businesses trying to enhance their customer experience are convenience, pricing, product experience, and trust. A company can build a loyal and satisfied customer base by offering more transparent pricing, competitive pricing, and building trust with customers.

After reading most of this book, you should now be better equipped to meet the buyer's requirements. You may have decided to extend your business hours, offer more affordable options, eliminate fine print and rules, surprise customers with additional perks, or provide a better experience by offering a 24-hour response time and increased guarantees. These are all excellent strategies to help you retain and attract more customers. However, it is essential to remember that your efforts must be constantly reviewed and updated as customers' demands and expectations continue to evolve.

As a business owner or leader, you must create a culture where all employees are committed to placing the customer at the center of the universe and working backward. This culture means that everyone in your organization should be aligned and assigned a purpose to constantly improve the customer experience. This customer-centric approach will help you build a loyal customer base and grow your business.

I will provide you with content and suggestions on various strategies that you can use to achieve your goals. However, it is essential to recognize that you must commit to the long-term and focus on sustainable solutions. One-time fixes may provide temporary relief, but more is needed to sustain your business in the long run. By adopting a customer-centric approach and investing in continuous improvement, you will be well positioned to succeed in today's competitive business environment.

As we conclude this chapter, it's evident that cultivating a customer-centric culture is more than just a good practice; it's a cornerstone of sustainable business growth. The insights and examples shared in this chapter testify to the transformative power of putting the customer first. Whether you're leading a multinational corporation or a small startup, the principles of customer obsession can propel your business to new levels of success. Remember, your customers are your most valuable asset in the journey toward organic growth. Building a culture that consistently prioritizes their needs and expectations lays the foundation for a thriving, resilient, and growth-oriented business.

Summary

Culture is a complex state encompassing shared values, beliefs, and behaviors that shape a particular group of people's identity and sense of belonging. It is a powerful force that can influence how individuals think and act and is often deeply ingrained in their daily lives.

In the business world, companies often create unique cultures based on specific goals and priorities. Some companies prioritize operational efficiency, focusing on streamlining processes and maximizing productivity. Others prioritize product-driven approaches, which means they place a strong emphasis on developing innovative products and services. Finally, some companies prioritize customer-centricity, putting the customer at the center of everything they do.

Regardless of the specific focus, a company's culture can be seen in every aspect of its operations, from its products and services to its customer service and internal practices and policies.

Operationally Efficient: Companies such as Walmart, UPS, and McDonald's prioritize operational efficiency, competitive pricing, and convenience to satisfy customers.

Product Driven: Companies such as Apple, LG, and Verizon focus on driving great product experiences with their innovative products. They have a culture centered around continuous innovation to attract customers.

Customer Intimate: Companies like Starbucks, Nordstrom, and Marriott prioritize the customer experience.

Not all organizations operate with a well-defined culture. Instead, they exist solely to make ends meet, hoping customers will choose to do business with them. These organizations can achieve growth and success with a clear strategy and accompanying culture. Unfortunately, many small businesses fall into this category.

While some companies claim to be "customer-centric," it's often just a buzzword to them. But a few genuinely embody a customer-centric culture, and Amazon is one of them. Amazon puts its customers at the forefront of everything they do and works backward from their needs. Their number one leadership principle is to obsess over the customer. Those who support a customer-centric culture believe that by taking exceptional care of the customer and fulfilling their needs, everything else will fall into place, including revenue, market share, and profits.

CHAPTER 13

Vision and Mission

Let me ask a strategy question. As a seller, is your mission to generate revenue and profit, or is it to satisfy buyers? There is a fundamental difference in those approaches. While they are related, the culture of the seller may vary differently.

When buyers consider purchasing, they evaluate potential options based on several critical criteria. These criteria vary based on individual preferences and values but generally include price, quality, reputation, and convenience. While these factors are important, personal values can also heavily influence buying decisions. It is essential to remember this when considering what to buy and from whom.

1. Convenience
    a. Saving Time
    b. Saving Effort
    c. Being Simple
    d. Being Easy
2. Pricing
    a. Competitively Priced
    b. Transparently Priced

3. Product experience
    a. Shopping experience
    b. Quality of the product
4. Trust

Dominant companies have a distinct advantage in the market as they meet all four criteria buyers use to make purchasing decisions. Meeting all buyer needs eliminates any excuses for buyers to purchase elsewhere and establishes a strong brand image for the company.

As buyers, we are all driven by our unique interests and values, our personal value system. We are all buyers, but not everyone is a seller.

When sellers shift their perspective from being buyers to sellers, they tend to prioritize their seller interests over the interests of their buyers. Focusing on sellers' needs rather than buyers' needs ultimately leads to a conflict of interest between buyers and sellers. It also produces a "Big Mystery," or lack of information or direction in business decisions, as sellers' shifted focus to their own interests can alienate buyers.

I'll use a football analogy—defenses game plan based on their knowledge and study about the opposing offense. If the opposing offense is run-oriented, they likely will game plan around that. Most pay attention to the offense and install the defense they like or are most comfortable playing that they believe has the best chance to stop the opposing offense.

So, knowing what buyers want, sellers should "game plan" for that and meet those needs, right? One would think so, but that is different from what often happens. Sellers shift their focus from what customers want to what they, the sellers, do well or want to do. It's what I call a big mystery. We are all buyers and have the same needs, but when we buyers become sellers, we shift our focus away from what buyers want to what we, as sellers, want and need.

Although revenue, margins, and profit are essential for any seller or company to succeed, it is equally crucial for them to focus on meeting the four criteria that buyers use to make their decisions. When sellers cater to buyers' needs, they realize it is the key to generating revenue, margins, and profit.

Meeting all four criteria is a challenging feat. For instance, most sellers would appreciate and enjoy having the purchasing power of retail giants like Walmart and Target, so they could compete on price while still maintaining their profit margins. However, the reality is that Walmart, Amazon, Target, and others have far superior pricing power over millions of small businesses. In such cases, sellers competing against Walmart should focus on the other three criteria: convenience, product experience, and trust. Sellers need to cater to the personal value systems of their buyers as much as possible to establish a solid and loyal customer base.

One advantage all businesses and sellers should have over large sellers like Walmart, Target, and Amazon is their people. Every company ultimately boils down to people. People make the difference. If you are a seller and don't have the buying power of larger organizations, you must deliver on the other three buying criteria in a bigger and better way. That all starts with your people and your culture. You must be more convenient, saving your buyers time and effort by making your processes simpler and easier. You must price as or more transparently than others, provide a better product experience, and be more trustworthy. As previously stated in an earlier chapter, you must constantly refine and improve on all of those. You can only do it with the right people and being aligned with the right culture.

As buyers and sellers, we are often driven by contrasting motivations. While buyers seek the best value for their money, sellers focus on their growth and success. However, sometimes sellers may forget what drives buying behavior and only consider their strengths as sellers. This narrow focus can cause them to lose sight of the buyer's perspective, which is crucial for success. Therefore, it is essential to maintain a balanced perspective as both a buyer and a seller to achieve mutual success.

Regarding buying and selling, it's common for both parties to be primarily concerned with their interests. However, sellers need to understand that by focusing on the needs and desires of the buyer, they can ultimately achieve growth and success in their business. By prioritizing the buyer's perspective and striving to meet their needs, sellers can build trust and loyalty, increasing sales and profits.

Regardless of whether it's a sports team or a workforce, every team has specific objectives that need to be accomplished. Ensuring every team member understands their role in achieving these objectives and is aligned with the mission is paramount. Rather than merely having a job, employees must have a clear sense of purpose. The purpose of every employee within a company should be to help the organization achieve its overall mission.

Simon Sinek professes that companies should have a mission and vision statement. The mission statement should reflect how you are to accomplish your objectives, and the vision statement reflects the why or your overall purpose. This is a strong sentiment, and I like this approach. But, because most companies, for whatever reason, believe they need a mission statement, I will focus on that.

Most people can be motivated to work towards a worthy cause, provided they clearly understand the overall mission they're striving to achieve. It's not merely about standard productivity metrics. It's about how every employee's job or function contributes to the company's mission.

As I mention the importance of mission and culture, allow me to elaborate on the typical mission statements that companies tend to use. Firstly, it's essential to question whether you and your colleagues know your company's mission statement. If you were to ask your colleagues to recite it randomly, how many could do so? My educated guess is that not more than 20% of them can.

A mission statement usually contains overused phrases such as "We will provide value to our customers, employees, and shareholders."

These statements seem hollow and lack any natural substance or value for employees. One becomes worthwhile if the employees are motivated and guided by a mission statement.

I advocate for a mission statement that every single seller can adopt. It's simple, to the point, effective, and applies to every individual in every company. Here it is:

We seek to reduce or eliminate buyers' excuses to purchase from us.

Buyers make excuses not to buy from a seller for the following reasons:

- The buying process or product is not convenient.
    - It takes too much time.
    - It takes too much effort.
    - It is too complex.
    - It is too hard.
- The price the seller wants isn't competitive.
- The pricing is confusing, and the customer isn't certain of the full cost.
- The product experience is not great.
- Others provide a better experience for the buying process and/or product.
- The customer doesn't believe what the seller is saying.

Running a successful business involves considering not only the needs of customers but also those of employees and shareholders. The key to generating revenue, promoting growth, and increasing shareholder value is prioritizing customer satisfaction and removing obstacles preventing them from choosing your brand over others.

Customers often make excuses for not buying from a particular seller, such as long wait times, complex processes, high prices, poor customer service, unclean facilities, unprofessionalism, lack of trust, unclear pricing, no guarantee, and poor warranty terms. Although some factors may not be under a seller's control, others can be eliminated or minimized.

To be a successful seller, you must focus on what you can control and work towards eliminating customers' excuses for not buying from you. Doing so can improve customer satisfaction, generate revenue, create a better employee working environment, and increase shareholder value. Magical things happen when employees understand their purpose rather than just their job. I believe that the best employees are those who feel they make an impact or make a difference. When challenged to eliminate buyer excuses, they now have a worthy mission and strive to contribute to the company's overall mission. Employees with no purpose likely feel they have just a job and are not making an impact. Great employees and teams want to win. Enable them to win by eliminating the excuses of the buyers and making the company grow consistently.

The company's mission is a fundamental goal that every employee can relate to and strive to achieve, regardless of their position. It is a purpose that is understandable and inspiring to all.

To illustrate this idea, let me provide specific examples of how every employee can contribute to fulfilling the company's mission. Employees can find their purpose and a meaningful role in the organization by doing so.

When selling a product or a service, the operations team should focus on ensuring efficiency and reducing waste. Any inefficiencies or wastage can add cost to the equation and may result in buyers sharing the costs, which is never desirable. The procurement team is crucial in buying the highest quality raw materials at the most competitive price possible so that buyers can get the desired experience.

The finance and accounting department is critical in providing the necessary metrics and Key Performance Indicators (KPIs) that the seller should use to determine if they are accomplishing their specific mission. Based on each employee's role, it should go down to the lowest level possible. Employees should know if they are winning or losing and contributing to their company's mission. Measuring the number of customers gained or lost and their reasons is essential. Customers are the lifeblood of a company, and their experience drives everything.

The product development team should focus on accommodating buyers' four criteria when purchasing. Product development aims to improve convenience by saving buyers' time, making the product offering more straightforward and accessible, and building customer trust.

Marketing is crucial in educating potential buyers and promoting the benefits of the products and the sellers. The brand promise must align with the buyers' criteria, and the marketing team should develop and hone their messaging to highlight these benefits.

The sales team's primary purpose is to convert prospects into revenue by meeting the four buyer's criteria and making it simple and easy for the buyer to say "yes." However, sales have a more critical purpose than just closing deals. They are also responsible for collecting and recording why buyers don't purchase. This intelligence is crucial for the organization to determine where to focus its attention and accomplish the mission of eliminating or minimizing excuses.

The IT team's primary purpose is to make everyone in the organization more efficient by reducing costs through automation, enhancing the buyer's experience, and setting priorities based on the impact on the buyer. IT projects often run the gamut, being all-encompassing, with some focused on internal efficiencies and others that do touch and impact the buyer or customer. In a customer-centric environment, it becomes easier to set the proper priorities with IT projects and place those that impact the customer at the top of the list.

Crafting a comprehensive plan for organizing, structuring, and measuring a company can be daunting. However, ensuring that each department and person has a specific role that aligns with the organization's mission is essential. By doing so, everyone will work in the same direction, and the company can achieve its objectives effectively. Therefore, it is crucial to keep these points at a high level while ensuring that every individual in the organization has a clear sense of purpose and direction.

Aligning an entire company towards delivering a truly customer-centric approach can be achieved by creating and managing customer-centric KPIs. A good starting point is to measure the number of customers, customers gained, customers lost, and lifetime value of a customer.

However, some businesses may need help to measure these metrics as they do not cater to repeat customers. For instance, convenience stores, restaurants, and retailers don't necessarily target or seek repeat customers. In such cases, creating and maintaining a loyalty program is an excellent way to measure customers, and it can also be used as a leverage point for marketing.

Measuring lost customers and lost opportunities is crucial for businesses with sales roles. Tracking why a customer left or didn't buy can help minimize excuses. Scorecards and KPIs around these metrics are vital.

Developing and documenting processes and measuring the time and effort a customer takes to engage with you is essential. Measuring how easy or difficult it is to engage with your company is crucial. Determine how simple engaging with your product and how it relates to convenience factors is. Conduct surveys with your buyers or prospects and turn their feedback into valuable KPIs. Net Promoter Scores are excellent KPIs that let you know how likely customers are to recommend you to others.

Create a customer team or committee whose mission and purpose are to evaluate various ways to measure buyer satisfaction and report on those as often as your financial reports. Turn your reports around from purely internal performance to external buyer information. At one point, pharmaceutical company Glaxo Smith Kline transitioned their sales bonus plan from top-line sales, or revenue, to customer satisfaction surveys. Just because customers are buying doesn't mean they are fully satisfied.

As we wrap up this chapter, we've unraveled "The Big Mystery" of the buyer-seller dynamic, revealing the importance of aligning with buyer criteria to achieve business success. This chapter has provided insights into your customers' minds, equipping you with the knowledge to cater to their needs effectively. Moving forward, remember that the key to unlocking organic growth lies in understanding and meeting these criteria. By focusing on convenience, price, product experience, and trust, you can build a business that meets and exceeds customer expectations, paving the way for sustainable growth and profitability.

Summary

As buyers, we all have unique criteria to evaluate the products and sources we purchase from. However, when a buyer becomes a seller, they often need to remember or adhere to the criteria they use as a buyer. Instead, they focus on priorities such as top-line revenue and profit. They need to recognize that buyers are precisely what drives revenue and profit. That juxtaposition is exactly what embodies the "Big Mystery."

Meeting all four of the buyer's criteria is incredibly challenging, and most companies must make some trade-offs to deliver. Moreover, each buyer has an evolving set of values, making it even more challenging to meet every need.

To foster a customer-centric culture, it is paramount to have alignment throughout the organization. Like a rowing team, reaching the goal line is inevitable if everyone paddles in perfect harmony and the right direction.

Many companies rely on a mission statement, but it rarely sets the stage for getting everyone aligned. A simple mission statement, such as "We seek to eliminate any excuses buyers may have for not purchasing from us," can provide clarity and purpose for every department and member of the organization.

All departments in any company play a role in minimizing the excuses buyers must purchase from a seller. When each employee understands their purpose as it is aligned to the company's overall mission, employees feel they make a difference and an impact and are more satisfied with what they do.

Operations: Ensure waste is eliminated so operating costs are the lowest and pricing can be lowered without negatively impacting margins.

Sales: Track why buyers are not buying and why buyers are leaving.

IT: Create processes and efficiencies that make the product experience better for the buyer.

Marketing: Create messaging that is directed to buyer needs. Manage the website so it saves buyers time navigating and engaging.

Accounting: Add customer-centric metrics that are meaningful but especially measure the number of customers, new buyers, lost buyers, and lifetime value of a customer.

When each role in the company can be tied back to the company's overall mission and are given some autonomy and authority to make the changes to improve the company's deliverables against their buying criteria, magical things happen in the way of growth.

CHAPTER 14

Innovation and R&D

In this chapter, we will uncover the engines of growth that propel businesses forward in today's fast-paced market. We will explore how research and development and a culture of innovation are crucial in meeting and surpassing customers' ever-evolving needs. From technological advancements in electric vehicles to the revolutionary impact of mobile technology, we will examine how businesses across various industries have leveraged innovation to stay ahead of the curve and foster organic growth.

Research and development are crucial for many companies, as they constantly develop and advance new products. When we think of R&D, we often associate it with technical industries or companies such as pharmaceutical companies that tirelessly work to create new medicines and therapeutics to keep their growth prospects healthy. Similarly, aerospace companies are constantly researching and developing materials and products to improve their technology and make it more efficient. Moreover, in the electric vehicle industry, R&D is focused on studying and enhancing batteries to offer drivers longer ranges and lower costs.

Innovation is the driving force behind R&D departments. It is the process of introducing new ideas, methods, products, or services that bring significant positive change and improvement in various areas of human endeavor. R&D departments create and apply novel concepts, processes, technologies, or a combination of these to address existing challenges, meet evolving needs, and capitalize on emerging opportunities. This innovative approach has led to many breakthroughs and advancements in various industries, making our lives easier, safer, and more convenient.

Innovation is a process that involves transforming and enhancing existing systems, products, or services or creating entirely new ones to drive progress and generate value. It requires pushing beyond conventional thinking and boundaries of knowledge to find novel solutions. Innovation can occur in various domains, including science, technology, business, social systems, arts, etc.

Although many believe innovation is only about new processes and inventions that enable us to accomplish what has never been done before, it also includes improving the convenience of existing products or services. Innovation can be as simple as reducing the time and effort required to perform a task or as complex as creating a new technology that transforms our lives.

For instance, the invention of electricity, the phone, and the airplane were all huge innovations that transformed our world. However, innovations such as artificial intelligence, the internet, and the computer have also significantly impacted our lives. Even adding camera features to cell phones was an innovation that improved our product experience and made our lives easier. Cameras were already in existence, as were phones and cell phones, but the innovation of combining them into a single device creates added convenience. It's innovation.

Innovation is about finding new ways to improve our lives, whether by simplifying complex processes or creating new technologies that enable us to accomplish more in less time. It's about making tasks easier, reducing complexity, and improving convenience. Innovation is a vital driver of progress that helps us push boundaries, find novel solutions, and create a better future.

Let's talk about the virtual assistants we have grown to love, such as Siri, Alexa, and Google. The ability of an object to understand and interpret our voice is amazing. These digital helpers are designed to save us time and effort. For example, instead of walking over to the light switch, we can ask Siri to turn on the lights when properly connected. These virtual assistants aim to provide an exceptional experience while saving us time and effort.

Innovation is essential to our lives as it enhances our way of living. When we talk about mass production, we mean using techniques, procedures, and technology to reduce costs and provide us with affordable prices. Imagine if we had to assemble a car without the help of robots. With manual tools like wrenches, screwdrivers, hammers, and drills, the costs would be unimaginable. Hence, innovation drives growth and makes production of complex machinery much more efficient.

Electric vehicles are an innovation. EVs are getting a lot of attention, and many are pushing hard for the United States to transition from combustion engines to electric, as the impact will have a very positive impact on the environment. With EVs becoming the standard, the carbon footprint from consuming and burning fossil fuels will be reduced. There should be a lot of positive sentiment for improving the environment.

There are far fewer moving parts in an electric vehicle, and thus, the maintenance cost is far less for an EV than for a combustion engine. I, for one, would like nothing better than to stick it to OPEC, the organization that controls much of the world 's supply and demand to prop up crude oil prices. A no-brainer, right? Is it good for the environment, and should we stick it to OPEC? Sign me up.

Not so fast, my friend. EVs, for now, cost significantly more than combustion vehicles. But I believe the big rub to gaining more adoption is the ability to have batteries that provide greater range and to be charged much faster than what is being offered today. Gas stations are abundant, and charging stations are not. I can fill up my gas tank in less than 10 minutes, and it's at least double that time to charge fast and get to 80% fully charged.

Thus, the convenience, or time and effort specifically, is not as strong as with combustion engines. While EVs have their advantages, they are contrary to our buying criteria for now. Another challenge is that almost every day, news articles are published about advances on the horizon for driving range and charging time. Does the average buyer want to buy at a premium today when technological advancement results in a product that will provide a greater range, more charging stations, and less time to charge than today? The obvious answer is no. When EVs are priced more competitively, are more convenient than combustion engines, and have experience equal to or better, then EVs will be more widely accepted by buyers.

Innovation is about being more convenient, providing a better product experience, and costing less. You can see innovation's impact everywhere, whether at home or work. It's vital to remember that innovation is only sometimes about introducing something new. Instead, innovation is often about improving existing products and services to make them more affordable, convenient, and user-friendly.

It's essential to differentiate between variation and innovation. When you visit a liquor store, you'll notice an increasing number of beer and seltzer brands and varieties. A new beer is not necessarily an innovation but a variation. However, a lemon seltzer adds a flavor that enhances the product experience. In that case, it's still a variation and not necessarily an innovation.

In 2020, approximately 4.4 million new business applications were filed in the United States. This could have resulted from the COVID-19 variant and the opportunities brought about due to the pandemic. Regardless, millions of new businesses start each year, yet 50% won't survive past year five. Many of these businesses started without any clear strategy, but just because someone wanted to become independent or further their passion. However, many start because they see a market opportunity where buyers are unsatisfied or do not meet their needs.

Imagine a bustling world of business where, every day, thousands of companies open their doors, each striving to outshine or, more harshly, kill its competitors. Even established companies can't afford to rest on their laurels in this highly competitive world or risk losing to new start-ups. Hence, every business must constantly improve, like a cross-country runner covering more distance daily to improve their time.

So, how can companies maintain this? Well, every company should have someone responsible for Research and Development. It can be something other than a dedicated department or a person; it can be split among team members. The person or team assigned to this task should identify and track their competitors, evaluating them according to the four buyer criteria. These criteria can help companies determine how convenient their competitors are in terms of the buying process and the product itself.

For instance, evaluating the time, effort, simplicity, and ease of buying, getting services, or using the product can help companies identify improvement areas. By comparing their company's convenience levels to those of their competitors, companies can develop strategies to stay ahead. So, in this fast-paced business world, every company must have a research and development team to constantly track and evaluate their competitors while striving for continuous improvement.

Understanding your competitors' pricing strategy and comparing it to your own is crucial. How competitive are their prices? Is their pricing structure transparent and easily understood? Do they offer a wide range of products and services? If so, how does their pricing compare across each category?

Another vital factor to consider is the product experience provided by your competitors. If they sell merchandise, how do their products' style, durability, and comfort compare to yours? If they operate facilities, how clean and well-organized are they? Is their staff friendly, professional, and easily identifiable with uniforms and name tags? If they provide a service, how responsive and efficient are they?

Guarantees and warranties are also important aspects to consider when evaluating your competitors. Do they offer a warranty for their products or services? What are the details of their warranty? Do they have references available to verify customer satisfaction? Do they use testimonials on their website? What is their return policy in case of unsatisfactory purchases or services?

There are several methods for gathering information about your competitors. You can purchase their products or services and assess them in detail. Alternatively, you can hire a third-party service to research and provide feedback on your competitors' products, services, and overall customer experience.

When buying Product X from an alternative source, how convenient is it to purchase from the alternative sellers?

1. How convenient is the actual service or product the competitor offers?
    1. How much time does it take to buy?
    2. How much time is involved to get the product?
    3. How simple and easy is the alternative?

How competitively priced is the other option when buying Product X from an alternative?

How transparently do they price?

1. Any price surprises?

2. When buying Product X from an alternative, how do you rate the product experience? Is the product durable, comfortable, efficient, and quality-made? Do they treat you well?
3. When buying Product X from an alternative, what makes you trust the alternative? Do they offer a guarantee, warranty, and easy returns? How did you learn of the alternative?

Assigning the responsibility of gathering competitive intelligence to the Marketing department could be highly beneficial. However, given that the market is continuously evolving, it is imperative to collect such intelligence constantly. As previously mentioned, each Sales team member should keep track of prospective buyer excuses. By combining intelligence from both Sales and Marketing, companies can prioritize which aspects of their offerings require innovation.

Research and Development (R&D) should explore technology and processes to enhance buyers' criteria. R&D can play a crucial role in evaluating artificial intelligence and serve as a central hub for all significant initiatives to determine their potential impact on customers.

All departments involved must be aligned and committed to continuous improvement. They are working towards meeting each buyer's criteria by offering more convenience, competitive pricing, excellent product experience, and trustworthiness.

Entrepreneurs need to note that the optimal opportunity to start or purchase a business is when you can leverage opportunities to satisfy buyers' criteria. If you plan to start a business, ensure you can provide a more convenient and competitively priced offering, deliver superior product experience, and be more trustworthy.

If you are an existing business looking to develop a new offering, ensure that you are capitalizing on delivering an unmet need, such as providing convenience, a lower cost, a better product experience, and being more trustworthy. Companies launching a new product or service frequently look inward and tell themselves they are good at something and should offer it. They often overlook what buyers are looking for, emphasizing more on what they are proficient at delivering.

Entrepreneurs considering starting or buying a business should take the time to identify their "secret sauce." Investors, whether they are friends and family, venture capitalists, or private equity firms, are more likely to invest in a business that has a unique selling proposition, particularly one that can be sustained or protected from competitors. While unique processes may be involved, technology is often the driving force that sets businesses apart from their competitors. This technology or process is usually implemented to enhance efficiency (lowering costs leads to more competitive pricing and leverage), offering better product experiences or providing greater convenience for buyers.

Discovering and developing your unique selling proposition and constantly improving it is important so competitors cannot easily find a gap to exploit. Continually refining your unique selling proposition ensures your business remains competitive and relevant in an ever-changing marketplace.

As we conclude this chapter, it's evident that innovation and R&D are not just corporate buzzwords but essential elements of any successful business strategy. This chapter has armed you with the understanding that to stay competitive and achieve organic growth, your business must continuously evolve, innovate, and adapt to changing market demands. Whether spearheading a startup or leading an established enterprise, the lessons from this chapter encourage you to embrace innovation, invest in research and development, and always keep the customer's evolving needs at the forefront of your business decisions.

Summary

Innovation is a broad concept encompassing more than just creating new products or services. While developing something entirely new is undoubtedly commendable, innovation can also be achieved by providing added convenience, reducing costs, enhancing product experience, or building trust. It is unnecessary to accomplish something as ambitious as sending a man to the Moon to be innovative. Even minor improvements, such as reducing processing time, simplifying a process, or finding ways to cut costs, can be considered innovative. Making a customer's experience more user-friendly is another innovation that should be considered.

Research and development (R&D) should be an essential function of every company. At the very least, companies should research their competitors' practices to identify how to offer greater convenience, better pricing, improved product experiences, and increased trustworthiness. Marketing departments are well suited to lead the charge in researching competitors and suggesting improvements in these areas.

Entrepreneurs seeking to purchase or start a new venture should find opportunities to provide added convenience, better pricing, improved product experiences, and trustworthiness. These are the enhancements that buyers are looking for.

When offering a new product, companies should not only evaluate their strengths; instead, they should try to understand the market gap in relation to the buyer's criteria and fill that gap. This approach will help companies stay ahead of the curve and foster more innovation.

CHAPTER 15

The Value Proposition and Power of Messaging

Quite often, companies are confused about the meaning and importance of a good value proposition. Many times, a seller's value proposition is their feature proposition. A value proposition is really all about the value the buyer is seeking and the seller's proposition to fulfill that value. Whereas feature propositions are all about the seller's point of view on the benefits or features they offer and believe are of value to the buyer. More often than not, there is a disconnect between what the buyer values and what the seller believes is the value they offer.

Many companies fail to express their value proposition. Instead, they opt to convey what they do or offer simply. Let me assure you that buyers are ***mildly*** interested in what you do. That's right, they really don't care what you do, they care about solving the problem or need they have. Yes, they must understand your product or service, but they are seeking to help themselves, and the better a seller is at conveying the value they provide that matches the buyer's values, the better.

We have extensively highlighted what buyers value. They value convenience, saving money, receiving a great experience, and trusting the seller. The more a seller can incorporate these values into their proposition and messaging, the more relevant they will be, and the more a buyer will relate to the seller.

Let me give you an example. Let's pretend I provide a delivery service in Kansas City. It wouldn't be unusual to have a website whose landing page states, "We operate dozens of trucks and deliver Kansas City." However, a better value proposition would be "We deliver Kansas City on time, damage-free, and at an economical rate." Or, "For 50 years, we have been delivering Kansas City with stellar on-time performance and delighting customers like no other."

A good value proposition should be the focus of one's website and on their landing page and then pave the way for the seller's messaging in every other form.

This chapter unveils the crucial role of effective communication in the business world. This chapter explores how a well-crafted, Unique Value Proposition can be a game-changer in differentiating your business in a crowded market. Drawing from the success stories of companies like FedEx and Geico, we will learn how the right messaging, aligning with key buyer criteria, can captivate your audience, build trust, and drive sales. Prepare to transform your approach to communication and see how powerful messaging can unlock new levels of organic growth for your business. The right messaging is key to both customer engagement and employees as well.

Messaging plays a vital role in a company's sales strategy. Messaging is more than just an elevator pitch or a description of what you do; it is the core of your value proposition. It is essential to have a clear and concise statement explaining how your product or service solves a problem, fills a need, or improves other solutions in the market. This statement is called the Unique Value Proposition (UVP) or Value Proposition. A well-crafted Value Proposition informs buyers of your offerings and articulates why they should choose your company's products or services over your competitors.

Your Value Proposition is the key to convincing potential buyers that your company offers something unique and valuable, and it often serves as a critical factor in determining whether they engage with you further or not. Value encompasses the four criteria that buyers use to determine what they buy and from whom. Therefore, the more criteria you can include in your Value Proposition, the better. From a buyer's perspective and internally, it is important to support your company's culture.

If you communicate the unique aspects of your culture internally, it will eventually retain steam and emphasis. But if you include it in your Value Proposition, and everyone in the organization knows and professes it, then you are constantly reinforcing it, and everyone adopts it. This communication helps to ensure that your company's culture is aligned with your sales strategy, leading to better results and a more engaged workforce.

Simon Sinek's book *Start with Why: How Great Leaders Inspire Everyone to Take Action™* delves into the difference between "what" businesses do versus "why" they do it. Sinek argues that true leaders inspire others by emphasizing the purpose and values that underlie their business rather than just the products or services they provide. In this way, they can create a connection with their audience that goes beyond the transactional.

One way this connection is created is through a company's brand promise. Rather than simply stating what a company does, a brand promise communicates the value it compellingly provides to its customers. For example, saying "we sell widgets" doesn't provide any useful information to a potential customer. However, saying, "Our widgets are economical, save our customers time, and provide a great experience, guaranteed," speaks directly to the needs and desires of widget buyers.

By communicating a clear brand promise, businesses can set the expectations of their customers, which is crucial in creating a positive buying experience. If a company can deliver on its promise and exceed customer expectations, it can create a loyal customer base that will continue to do business with them in the future.

Fulfilling the brand promise is of utmost importance for a seller as it builds customer trust. A promise must be fulfilled to result in a good product experience; failure to do so can lead to negative feedback and a loss of potential future sales. Therefore, the entire organization must fully comprehend and embrace the brand promise and the value proposition to deliver on it effectively. This brand promise should be continuous, as improvement is always necessary.

A website is the primary platform for promoting a brand promise or value proposition. The value proposition should be strategically placed on the landing page to catch the buyer's attention and instill confidence. In today's digital age, buyers visit a seller's website with a specific need. Therefore, the website must reassure them that their criteria will be met. A clear and concise brand promise or value proposition displayed on the landing page can significantly impact the buyer's decision, setting the foundation for a positive product experience and a long-lasting relationship.

As a seller with a website, you have somewhere between four to five seconds to capture a buyer's imagination. No, they aren't going to buy simply by being awed by your landing page and value proposition. Rather, that value proposition will determine if the prospective buyer is intrigued enough to want to learn more and scroll down.

Many buyers are likely searching for a solution that better meets the needs they have. Their current solution could be taking up too much of their time and effort or is too complex and challenging to use. In some cases, the current provider may have raised prices or added non-transparent costs, making it more expensive than it used to be. The product experience may be worse than it once was, or the buyer may have been burned in the past without any guarantees or warranties to protect them.

Buyers often search for products under relevant categories online to find better alternatives. However, most search engines direct them to sellers who have paid to be listed at the top of the search results or are popular based on other searches.

When a buyer clicks on a website, they hope the landing page will provide clear information about the seller's offers. However, if the page only states, "We sell widgets," the buyer may not be intrigued enough to spend their valuable time researching the seller further. Most buyers spend four to five seconds on a website and may move on to the next option if they are not intrigued enough to learn more. Simply stating what a seller does is not sufficient. Prospective buyers really don't care what you *do*, they care about how you can *solve their problems*. businesses that successfully merge what they do with how they can solve a buyer's problems are more successful than those that do not.

Many websites do not convey their purpose effectively, and some use nonsensical language that fails to communicate what they are selling or why buyers should be interested. Having worked in the transportation and logistics industry for years, I can attest that the industry needs to articulate its value proposition clearly.

Recently, I came across a website from a last-mile delivery company that only had a landing page with the text "Now Hiring Drivers." Prospective buyers don't care if you are hiring drivers. In fact, maybe that statement will scare a prospective buyer off. Do they not have enough drivers to deliver my products on time?

On another website, the only information provided was the company's name, and the year it was established. But can we trust them because they've been around for a while? Will they solve our problem?

Sellers need to get their messaging right by clearly stating what they offer and how they solve a problem for the buyer based on their four criteria. Good messaging or value propositions that resonate with buyers can be found, but it's essential to be aware of what makes bad messaging and what not to do when crafting yours.

Since I have been in transportation and logistics, do you recall this Value Proposition?

When it absolutely, positively has to be there overnight.

The company's Value Proposition was a compelling message that left a lasting impression on anyone who saw it. It was visually represented by a striking image of a large cargo airplane adorned with the company's distinct colors and branding. The image alone was enough to convey that the business was in the air cargo industry, but the message was much more profound.

The Value Proposition focused on two key elements: time and trust. The promise of overnight delivery ensured that customers could rely on the company to meet their deadlines and deliver their goods on time. The phrase "absolutely, positively" acted like a guarantee, instilling confidence in customers and making them feel assured that their cargo transport service was safe and trustworthy.

This powerful Value Proposition drove the company's success, and everyone in the organization embraced it. From the CEO to the front-line staff, everyone understood the importance of the messaging and was committed to delivering on the Brand Promise.

Here is one of my favorite value propositions I discovered at the Lake of the Ozarks in mid-Missouri. I had to stop and take a picture. Along the side of the road, there was a small boat completely encapsulated by a red boat cover, and it said:

Uncover a Clean Boat in Less than 60 Seconds!

As an avid lake-goer, my top priority is to hit the water as soon as possible. That's why the idea of a boat cover that can be removed in just 60 seconds caught my attention. Spending time cleaning the boat is different from my idea of fun, and I'd rather be out on the water enjoying the scenery. I came across a business specializing in boat covers, but they were selling convenience. Their product not only saves time by allowing for a quick 60-second uncovering but also saves effort by eliminating the need for boat cleaning.

You have likely heard this one:

> ***Fifteen minutes could save you 15% on car insurance. Geico, get a quote.***

Geico's current advertising message is crafted to appeal to those looking for a cost-effective, time-saving solution to their car insurance needs. If Geico's slogan were "2 hours could save you 15% on car insurance," it would likely not motivate potential customers to contact them. However, by advertising that it only takes 15 minutes to get a quote, Geico is addressing the needs of cost- and time-conscious people.

Their message communicates that they offer car insurance while highlighting their service's benefits. By spending just 15 minutes, a person could save $75 on a car insurance policy that costs $500. This straightforward and simple message makes it easy for potential customers to understand the value of Geico's offerings and motivates them to act.

Another one:

Dollar Shave Club.

> ***Our Blades are f***ing great. Shave Time. Shave Money***

The company's name, Dollar Shave Club, tells us what they do. They get our attention with the first sentence but then hit us with a clever convenience factor and pricing consideration.

Here are a couple of examples of a lesser-known company's messages:

Vaerus Aviation

> ***Aircraft Ownership is Complicated Simple. We ensure the only thing our clients have to take care of is their drive to the airport.***

The company's name may imply that it deals with aviation, but its focus goes beyond that. Their solution aims to simplify the challenges and intricacies associated with owning an aircraft, which can be overwhelming and time-consuming. Whether you are an individual aircraft owner or manage a fleet, their service provides a comprehensive solution to make the process easier and more convenient, ultimately saving you time and resources. Their expertise and attention to detail make aircraft ownership a hassle-free experience.

MIB

> ***Peace of Mind Every Step of the Way, Experience the MIB Way. Never wait, Professional Chauffeurs. Sanitized Vehicles.***

Despite its obscure name, this company prioritizes product experience. However, it takes time to be clear what services they offer from the first part of their message. Only through the sub-heading we learn they specialize in providing a luxurious chauffeur and car service. Notably, on the lower section of their website landing page, they proudly declare the following:

> *Early and On-time Pickup (Time-related benefit)*
>
> *No Hidden Fees (Transparent Pricing)*
>
> *Sanitized Vehicles (Product experience)*

The examples of messaging above demonstrate how each of them successfully promotes some aspect of meeting the buyer's criteria. While some examples promote only one solution of the four criteria, they primarily lead with that solution and convey their message through visual association, name, or statement. These examples appeal to buyers' requirements to determine what to buy and from whom.

A messaging strategy that incorporates all four criteria is optional. Adding all four criteria can make the messaging sound clumsy. For example, a message that states, "Our widgets save you time, are competitively priced while delivering you a great experience, guaranteed!" hits on all four primary criteria. But it isn't very concise and almost sounds clumsy. The most important aspect of a messaging strategy is that the message should be genuine and delivered on your brand promise. The messaging should not contain any inaccuracies. Getting your messaging right is not easy, but when done well, it will resonate with buyers and potentially become your mantra.

Some companies use descriptors in their messaging that are effective and hard to challenge. For instance, stating that a company is "Number 1", "The Best," or "Voted the Favorite" are great examples of methods to build trust. However, these claims are not necessarily verifiable. Nevertheless, such messaging can be effective compared to other companies that do not exude trust in their messaging.

There are also Unique Value Propositions that keep the buyer from knowing what they do for them.

Intel uses "Intel Inside." Maybe they are conveying a great product experience with quality, but it's not explicit enough.

Got Milk? A slogan that only conveys a little relative to the four criteria we buyers use.

Several generic value propositions are weak because they do not meet our buying criteria. Here are examples:

- “We Do Things Differently”
- “Quality You Can Trust”
- We’re Here for You”
- “The Future of____”
- “Simply The Best”

Since a seller’s website is so important, you should evaluate your current value proposition. Make sure it states what you do but, more importantly, how you are solving the problems of any potential buyer. This is not an easy task and may require multiple iterations to get it right.

As we conclude this chapter, the importance of messaging in the business landscape has never been clearer. You are now equipped with insights into creating a compelling Unique Value Proposition that resonates with your target audience.

Remember, effective messaging is more than words; it’s about connecting with your customers’ needs and values, establishing trust, and setting the foundation for lasting relationships. Whether you’re a startup or a seasoned enterprise, refining your messaging strategy is a step towards distinguishing your brand and accelerating your journey towards organic growth.

Summary

When selling products or services, messaging is a vital aspect to which every seller should pay close attention. Unfortunately, many sellers fail to convey a message that effectively captures the attention of potential buyers. Rather than highlighting why their products/services are the best solution to buyers’ needs, they usually focus on the features or benefits they offer. However, the most compelling messaging should include at least one of the four critical criteria that buyers value. These criteria include convenience, competitive and transparent pricing, product experience, and trustworthiness. By conveying how your product/service offers any of these criteria, you stand a better chance of making a lasting impression on your potential buyer.

It's worth noting that trying to include all four criteria in your messaging can make it sound unappealing and confusing. Therefore, focusing on one or two criteria that align with your target audience's needs is crucial. Focus on the criteria where you can most differentiate your business in your messaging; hopefully, those that most align with your client's needs. Moreover, accuracy is essential—your messaging should be YOUR brand promise, and your buyers expect you to deliver on it.

It's not only potential buyers interested in messaging; your employees care about it, too. Your messaging should be the driving force that aligns with your company's culture and values, and your employees should be driven to adopt and deliver on the promise.

CHAPTER 16

Take Action

In 1975, an engineer at Kodak named Steven Sasson invented the first digital camera. Sasson's invention was revolutionary. The prototype he built was about the size of a toaster and took black-and-white images at a resolution of 0.01 megapixels, storing them on a cassette tape. The camera took 23 seconds to record a single image.

Despite this groundbreaking innovation, Kodak's management was not enthusiastic. The company's leadership was deeply invested in the highly profitable film business and was concerned that digital photography would undermine its core business. Their reluctance was driven by several factors:

1. **Profitability of Film**: At the time, Kodak had a dominant position in the film market, which was immensely profitable. The company was earning significant revenues from selling film, paper, and chemicals used in the development process.
2. **Fear of Cannibalization**: There was a fear that digital technology would cannibalize their existing film business. This concern was not unfounded, as digital photography would eventually render film obsolete.
3. **Underestimation of Digital Technology**: Kodak underestimated the pace at which digital photography would improve and become mainstream. Initially, digital cameras were seen as inferior in quality compared to film.

Missed Opportunities and Strategic Failures

1. **Lack of Investment in Digital**: While Kodak did dabble in digital technology, they didn't invest sufficiently to become a leader in the market. Competitors like Sony and Canon took the lead in digital cameras.
2. **Late Transition**: Kodak eventually embraced digital technology, but by the time they did, it was too late. The market had already been captured by other players who had been quicker to adapt.
3. **Failure to Innovate Business Model**: Kodak could have used its early lead in digital technology to innovate and create new business models around digital photography. Instead, they clung to the old model centered around film.

Consequences

By the time Kodak attempted to shift its focus to digital, the market dynamics had changed drastically. The company filed for bankruptcy in 2012. Although they have since emerged from bankruptcy and shifted their focus to other areas like printing and imaging solutions, Kodak's story remains a cautionary tale.

Lessons Learned

1. **Embrace Innovation**: Companies need to embrace disruptive innovations, even if it means cannibalizing their existing products.
2. **Adaptability**: Businesses must be willing to adapt their strategies to changing market conditions and technological advancements.
3. **Forward Thinking**: Organizations should foster a culture of forward-thinking and not be overly reliant on past successes.

Kodak's downfall was not due to a lack of technological capability but rather a failure in strategic vision and adaptability. Their story highlights the importance of being open to change and not being blinded by short-term profitability at the expense of long-term sustainability. Kodak failed to take action.

You are a successful seller. You are generating revenue, making a profit, and experiencing growth. All good! Right? You tell yourself you need to stay on course! There are reasons to stay the course if you are successful, but also reasons to be paranoid. Good companies like Blockbuster, Kodak, Borders Books and Sears had flourishing businesses but failed to act or react to changing trends, technology, and competitors.

This chapter is my call to arms for businesses seeking to thrive in a competitive market. In it, I emphasize the importance of continuous improvement and staying aligned with the evolving needs of your buyers. From process optimization to the strategic use of AI, we'll explore practical steps you can take to refine your operations, enhance the customer experience, and ensure your products and services consistently meet the four critical buyer criteria. Get ready to transform your approach to business and embrace the tools and strategies that will drive your growth.

After conducting thorough research, I've confirmed the four crucial criteria buyers consider when purchasing. It's important to note that these criteria are not the only factors influencing a buyer's decision. The constantly striving for improvement culture is also essential to ensure long-term growth. One cannot rest. It's a bit like a game of golf. There is no way to shoot a perfect round, and the goal is to improve and do better constantly.

Recognizing that competitors will always look for opportunities to improve upon your shortcomings is vital. They may be existing competitors or new startups seeing an opening in the market. It's imperative to continuously improve your product to avoid losing customers to these opportunistic rivals. Competitors are always looking for ways to erode your market share, and if you let them beat you by improving upon the four buyers' criteria, then you deserve the outcome. Do not rest.

As such, it's essential to remember that the buyer's engagement is just as necessary as your product meeting the four criteria. The purchasing process should be simple, easy, and convenient for the buyer. If the product meets the buyer's standards, but the purchase process is a hassle, the buyer will likely look for an alternative source. You can have the greatest product on the market, but if the buying process you support isn't convenient and results in a great experience, you won't optimize your market share.

In conclusion, to stay ahead in the market, it's crucial to continuously improve your product and ensure a seamless purchasing experience for the buyer. Continuous improvement will ensure long-term growth and help you stay ahead of the competition.

In early 2023, I was appointed as a chair for Vistage, the world's leading peer group organization for business leaders. As part of my training, I joined a Vistage group as a member for six months to gain more knowledge and skills. Vistage is a unique organization that requires members to work one full day a month on their business rather than in their business. Getting out of the day-to-day once a month allows a leader to focus on strategies and long-term initiatives.

During our meetings, we devote almost half the day to continuing leadership education, which involves conducting interactive workshops on various topics. In my first workshop, we learned about process triage, identifying and prioritizing broken processes as entities expand and grow.

The speaker used a process map taped to a wall, roughly 30 feet, detailing every action and process from marketing to cash collection. We learned that it is essential to have metrics in place to determine the effectiveness of critical processes and identify areas that need improvement.

It was interesting to note, as I discussed with other leaders in the group, that most companies only map out a few processes instead of the entire set of processes used by the company, from marketing to invoicing and collecting payments. This realization highlighted the need for businesses to take a more holistic approach to process mapping to ensure optimal efficiency and productivity.

If you are looking to improve your business processes and build better relationships with your customers, there are a few key steps you can take. The first step is to create a detailed map of your existing processes, focusing on touchpoints where customers interact with your business. This analysis could include anything from initial inquiries about your products to the final stages of the purchase process, post-sale support, and follow-up.

As you map out your processes, it's important to go into detail to understand the customer's requirements at each step thoroughly. Mapping processes will help you identify areas where you can simplify the process or make it more convenient for the customer. You can also use this information to determine the amount of time, effort, and convenience factors involved in the process and any areas where there may be waste or customer frustration.

Once you have mapped out your processes, evaluating them from the customer's perspective is essential. This evaluation means getting customer feedback on how convenient and transparent your buying process is compared to other options. You can also ask them about pricing and how competitive your products are compared to alternatives. Finally, you can ask for feedback on the product experience and use this information to make improvements that better meet their needs.

By taking these steps, you can build trust with your customers and ensure that they are satisfied with your business experience.

Customer satisfaction, in turn, will help to improve your bottom line and strengthen your position in the market.

Conducting thorough research on your buyers' needs and preferences is crucial for the success of your business. Understanding your buyer's needs should take precedence over identifying gaps in your processes, though doing both will undoubtedly provide valuable insights into ways to serve your buyers better.

If you cannot meet all your buyer's requirements, it's essential to promote and leverage the ones you can. For instance, if you cannot offer competitive pricing, provide unparalleled convenience, product experience, and trustworthiness. Share the reasoning behind your offerings with your buyers, not just the products themselves.

To gather this information efficiently, set up a routine for collecting data from your buyers. Contact new buyers immediately and establish a regular cadence with repeat customers to ensure you take advantage of all opportunities. This approach will allow you to stay updated with the latest market developments, including new alternatives and why they succeed or fail. It's also valuable to reward buyers for sharing this information with you, as it provides crucial insights that are more valuable than traditional marketing efforts.

In today's digital age, having a website is a prerequisite for any business. It serves as your online identity and is often the first point of contact for potential buyers. Therefore, it is crucial to have a website that is not only functional but also visually appealing and professional. Having a high-quality website means investing in quality design and graphics that reflect your brand and instill trust in your potential buyers.

However, your website is more than just a digital storefront. It is your opportunity to communicate your value proposition and differentiate yourself from your competitors. The key is to focus on delivering your brand promise and addressing the problem your buyers want to solve.

Messaging about solving buyer's problems can be achieved through clear and concise messaging that resonates with your target audience.

Include a call-to-action (CTA) icon on your landing page to optimize your website further. This CTA should be easily noticeable and straightforward, allowing the buyer to act with minimal effort. Examples of effective CTAs include "Call now," "Book an appointment," "Learn more," "Buy now," or "Talk to a representative." By providing a straightforward way for buyers to engage with your business, you increase your chances of converting them into loyal customers.

It is important to ensure that your website has a prominently displayed "Contact Us" button in the menu. However, having someone accountable for responding to customer queries is equally crucial. Failure to do so can negatively impact the customer experience, ultimately leading to lost potential sales. Therefore, assigning a responsible person to respond to customer inquiries promptly and accurately is necessary. From personal experience, I have found that only about 50% of sellers respond via the "Contact Us" button when I choose to contact them.

To ensure that the website provides a seamless and user-friendly experience, it must be engineered in a way that is easy to navigate. Too many pages or options can overwhelm buyers and lead to frustration. Thus, a website model consisting of only a few sections or pages is recommended, with the landing page featuring a long scrollable page. This design approach will make it easier for buyers to browse your site, locate what they need, and easily purchase.

Building trust with potential customers is vital for the success of any business. One way to build trust is by displaying at least three testimonials from actual sellers below the landing page. These testimonials will serve as social proof and assure potential customers that your brand is trustworthy and keeps on its promise. Additionally, providing a high-level process of how the buying process works and what buyers can expect regarding delivery timelines is critical. This information will help buyers understand how simple it is to purchase on your website, thus increasing their confidence in your brand.

Creating a visually appealing presentation of your product or service is important when designing your website. If your website offers multiple tiers of pricing, make sure to showcase this information as well. Also, be mindful that your website has optimized navigation, so it's essential to provide a clear "call to action" button such as "get a quote now" to encourage potential customers to take the next step.

Sharing success stories and brief case studies on your website is essential to establish credibility and build customer trust. Such testimonials give customers real-life examples of how your product or service has helped others, making them more likely to trust your brand. Consider including a short story about your company's mission if it's compelling enough to capture the reader's attention.

Make sure that your website includes a transparent call-to-action icon for buyers to make a purchase or take another desired action. Your call to action should be prominently displayed at the bottom of the page so customers can access it easily.

If you run an e-commerce site, using a platform with a powerful and flexible search engine can help potential buyers find what they're looking for quickly. Highlighting free shipping and fulfillment time can also help persuade customers to purchase. Additionally, ensure you capture and share product reviews for each item, which can help build trust with potential customers. Finally, capturing email and/or cell phone numbers can inform customers about the status of their orders and estimated delivery time.

Chatbots have emerged as a popular customer service channel that allows businesses to accommodate customers who prefer not to use the phone. However, it is essential to ensure that the chatbot can answer customers' questions and provide quick solutions to their queries. To avoid frustrating your customers, you must limit the bot's functionality and provide prompt access to a human representative whenever necessary.

Phone trees can help direct customers to the correct department but can also be frustrating if they take too long to navigate. If your phone tree does not allow you to speak to a human representative by pressing "0," it may lead to even more frustration for your customers. While chatbots and phone trees are meant to lower costs for the seller, they should also enhance the buying experience for the customer.

Prioritizing the buyer's perspective when designing processes and technologies is crucial. Avoid installing systems that make your operations more efficient but negatively impact the customer's product experience.

Internally, it is advisable to set up meaningful customer-centric metrics and KPIs to evaluate your business's performance. Start by measuring the number of new customers acquired, the reasons for their acquisition or loss, and the total number of customers you currently have. Work backward from there to revenue, profits, and other metrics.

To gauge the effectiveness of your customer service, measure the number of "contact us" inquiries and evaluate the effectiveness of your loyalty programs. Determine program effectiveness by comparing the purchasing behavior of loyalty program customers with non-loyalty program customers.

To ensure that your company meets all the buying criteria critical to customers, create a customer committee responsible for collecting research data from buyers and evaluating processes and trends. This committee ensures that your company is meeting the needs of its customers and providing them with an excellent buying experience.

Artificial Intelligence, or AI, is a complex field that has taken the world by storm. Although many people are unaware, machine learning has existed for some time. You may have heard of generative AI tools, such as ChatGPT, that help with content creation, but there is another side to AI that is just as powerful.

Generative AI utilizes machine learning and Large Language Models (LLMs) to read and create content, including emails, and even develop software code. It constantly learns from data on the internet to improve its abilities, making it a valuable tool for sellers.

On the other hand, there is Applied AI, which allows businesses to use their data to make better decisions and predict outcomes. By merging their data with online sources, companies can gain insights that set them apart.

For instance, AI can assist with logistics decisions, such as determining the optimal location for distribution centers and shipping optimization. The best part is that the initial analysis is quick and efficient, and with the right inputs and algorithms, it's easy to rerun data in real time to make informed decisions.

AI is also utilized for forecasting, providing quick and real-time updates to forecasts when given the right input and programming.

One of the most remarkable features of AI is its ability to predict outcomes, such as when customers may leave. If you are a seller and can predict when a buyer might slow their buying and leave you as a customer, you can act much faster. Buyers will react positively when you can proactively address them and determine how you can better solve their problems. You can ask how you can make their lives more convenient, save money, improve their product experience, and be more trustworthy. Buyers will take that as a sign you care.

E-commerce businesses use AI to provide customized buying suggestions, making the checkout process more efficient. Suggesting other items to buy enhances revenue opportunities for sellers and makes it more convenient for buyers.

While AI is still in its early stages, it is increasingly used to reduce costs and improve quality. Robots, directed by AI, can accomplish much while lowering operating costs. Automation and AI are significant for sellers striving to lower costs and remain competitive.

Although AI can power chatbots, phone trees, and other customer-facing activities, businesses should remain cautious and avoid going overboard. Buyers are impatient, and their time is valuable, so excessive use of such tools may frustrate them.

AI is undoubtedly beneficial for helping businesses achieve their goals, and every company should research how AI can help them accomplish their mission.

Be mindful that customer-centric companies prioritize the customer and work backward.

As we close Chapter 16, I leave you with the tools and insights necessary for tangible action. The journey toward organic growth requires a commitment to constant improvement, a keen understanding of your customer's needs, and the strategic use of technology. By focusing on these areas, your business can rise above the competition, foster customer loyalty, and achieve sustainable growth. Remember, the path to success is an ongoing learning, adapting, and evolving process. Your willingness to take action today will define your business's success tomorrow.

Summary

To ensure that your business provides convenience, competitive pricing, transparency, an excellent product experience, and trustworthiness, it is essential to evaluate your processes, particularly those that interact with the buyer. Even if your product meets all four criteria, a complicated and time-consuming buying process can frustrate buyers and discourage them from executing the purchase.

Creating a detailed map of your processes is crucial to identify areas that require improvement. You should also communicate with your customers and gather feedback on how they rate your business concerning the four criteria, especially when they have other alternatives.

Your website is a critical component of your business. It is the first point of contact for many customers and can make or break a sale. Therefore, it is essential to have a clear message that delivers a solid call-to-action, testimonials, and conveys trustworthiness to potential buyers.

Finally, with the advancement of AI, companies can operate in new and innovative ways. By leveraging Generative AI and Applied AI, businesses can enhance their operations and meet the customer's buying criteria more effectively. You can explore new possibilities and ask yourself, "What if" you had access to certain information to deliver better products and services to your customers?

Most importantly, recognize and realize you will never be satisfied with the buyers' criteria. We buyers are always looking to save more time, exert less effort, and make life simpler and easier. We are always looking for deals and bargains, we always want a better product experience, and we want to trust the product and seller. As new technologies evolve, understand those that can improve upon what you deliver to your buyers, either via the product itself or the buying experience.

When you innovate and deliver more to the buyer, you will be rewarded with unstoppable growth.

CHAPTER 17

Conclusion

This book is a roadmap for businesses aiming to excel in a competitive marketplace. This final chapter delves into aligning business strategies with buyer preferences, emphasizing the importance of convenience, price, product experience, and trust. Drawing from a rich tapestry of examples, we explore how focusing on buyer needs can propel a business toward unprecedented growth and success.

It's common sense: give buyers what they seek, and you will win business. You will create advocates and repeat customers. You will grow and do so without dramatically having to increase sales staff and marketing budgets.

Buyers, be they everyday consumers or businesses, want the same thing. We want convenience. We want to save time. We want to exert less energy and effort. We want the problem made simple and the hard-earned easy. We buyers want competitive and transparent prices. We don't like surprises. We prefer to avoid overpaying. We want a great product experience. We want to feel as if we are unique. Please pay attention to us. We want to trust you. We hate scams, and we want assurance that you are fulfilling your brand promise and delivering what we naturally expect.

Do all of this, and you will grow. Yes, sellers must make a profit. As buyers, we don't care so much that you make a profit. Your job is to figure that out while providing us with everything we want. You will have challengers and competitors. A culture that supports continuous improvement and constantly improves upon what buyers wish will be the best way to ensure competitors don't swoop in.

In 1999, I embarked on an entrepreneurial journey by establishing my first significant company, freightPro.com. The advent of the Internet and the World Wide Web was a game-changer. I was inspired to leverage this new technology by making a Transportation Management System (TMS) accessible online.

It was an exciting time as venture capitalists in Silicon Valley were on the lookout for promising dot coms to finance, and I managed to obtain venture capital to accelerate the growth of my company. With the TMS, we aimed to simplify the complex and time-consuming process of LTL (Less than Truckload) shipping, which was much more intricate than shipping small packages with UPS or FedEx.

LTL carriers utilized intricate and proprietary rate tariffs. They offered substantial discounts to attract business, making it difficult to determine the base rate to apply the carrier's discount without a TMS. The discounts ranged between 70% to 80%, and choosing the commodity classification of the product was a Herculean task. We relied on a catalog nearly the size of an old-school Sears catalog to index almost every product and determine which of the 18 classes the product belonged to.

Back then, TMSs were not cloud-based, and they required hardware such as servers and incurred numerous ancillary expenses to operate correctly. We had to set up carriers for rate shopping and integrations for booking, tracking, and reporting on shipments, which was also time-consuming. Despite the challenges, we were determined to succeed, and our success paved the way for cloud-based TMSs that are now widely used.

When it comes to shipping goods, there are often numerous variables to consider, such as the origin and destination zip codes, the transit time, and the cost. However, determining the best carrier to use can be challenging, especially when there are exceptions to the discount based on whether the LTL carrier serves the points directly or works with a partner.

To make this process more convenient for customers, we developed a Transportation Management System (TMS) accessible through a browser with credentials. This system saves shippers money by eliminating the need to invest in their software. We loaded their rates and transit times and applied rules for discounts so every shipper could get the net rate for multiple carriers. After choosing their preferred carrier, shippers could book the shipment and track it until it was delivered.

One challenge with shipping goods is the complex commodity classification system. We simplified this by using density as the variable to rate the goods. We made it easier for shippers to measure and weigh each shipment and insert that information into our system. We then translated this information into density and negotiated with LTL carriers to convert this density into their preferred classification system. We also arranged a liability limit in case of loss or damage.

Shippers could see multiple carrier options and decide based on cost and/or transit time. Displaying all options was a more convenient, cost-effective, and user-friendly experience, which helped us to build trust with our customers. To further strengthen this trust, we allowed any shipper to book their first shipment under our credit terms.

Overall, this is just one example of how we used technology to simplify a previously very complicated existing industry and provide a better experience for our customers.

Amazon is another excellent example. Jeff Bezos understood how leveraging the Internet could enable him to provide a more convenient, less costly, and better experience for buyers. Yes, he started selling books but kept working on the model to what it is today.

Amazon adopted the number one Leadership Principle centered around a customer-centric culture. This principle goes beyond mere words and involves implementing processes and measurements to ensure you are truly obsessed with the customer.

To adopt this principle, start with the customer and work your way back. Focusing first on the buyer means making the buyer the focal point of your business and providing them with what they want. Customers desire convenience, simplicity, ease, and wish to save as much time and effort as possible. They also expect competitive and transparent pricing and an exceptional product experience. Most importantly, they want to trust the products and sellers they buy from.

To maintain this approach, everyone in your team must be onboard and working towards the same goal. As a seller, you also need to make a profit. However, distinguish competitive pricing from breaking even or losing money. Buyers will return to your business if you fulfill the four criteria of providing convenience, experience, trust, and competitive pricing.

Providing convenience, experience, and trust is under your control as a seller. While competitive pricing may only be partially controllable, finding ways to lower costs is vital without sacrificing the other three criteria. Offering convenience, experience, or trust for lower pricing is not a sustainable approach in the long run.

The ability to satisfy all four buyer criteria is undoubtedly a difficult task. If it were an easy feat, every business would be doing it, making it hard for a single company to stand out. However, if you can deliver on all four criteria, you can constantly attract new and repeat buyers. It is essential to note that social media is a potent weapon in this digital age. As I mentioned earlier, buyers seek recommendations from their peers before purchasing. With so many social media platforms available, it is easier for buyers to get information on a seller's overall experience. Unfortunately, negative reviews are more common than positive ones when things are unplanned. Buyers are often motivated to seek retribution, and they go to great lengths to do so. As a seller, this can be detrimental to your business.

As a seller, it is crucial to maintain an excellent online reputation. It would be best if you did your job well, and your reputation will eventually speak for itself. It is vital to understand that we are all buyers at some point. As a seller, you should avoid the trap of forcing a trade-off. Some sellers may be tempted to ask, "If I provide an excellent experience, will they pay for it?" or "If I'm the most convenient option, can I charge more?" This mindset significantly contributes to why many businesses fail to dominate their market.

Every seller is also a buyer, but they tend to prioritize their interests as a seller over those as a buyer. This self-interest is often evident when sellers force self-checkout on buyers despite it being an unpleasant experience for many. Unfortunately, this is just one example of a common phenomenon.

Remembering the cheese metaphor as a seller is crucial: provide options for your buyers. Offer a primary option like a block of cheese to appeal to those who prioritize low prices above all else. However, your options should also include more premium tiers like shredded or sliced cheese, which offer greater convenience or a better product experience. By doing so, you cater to buyers with different priorities and values instead of limiting yourself to a single option. Refrain from letting your self-interest as a seller determine your potential as a business.

To build and maintain a thriving business, it is essential to prioritize the satisfaction of your customers and create a culture that centers around them. To achieve this, it is recommended that you establish key performance indicators (KPIs) and measurements that revolve around the customer. Customer KPIs can include tracking the number of new customers gained and lost and understanding the reasons behind these changes. It is also essential to have your sales team record accurate excuses used by buyers who ultimately do not purchase from you. By analyzing this data using a Pareto chart, you can identify the root cause of the most significant loss of sales and take steps to address it.

Your messaging should go beyond simply describing what you do and instead focus on how you can help your buyers solve their problems. Everyone in your organization must understand and promote this message.

To stay ahead of the competition, it is important to maintain regular communication with your buyers and actively seek their feedback. Talking and learning from buyers will help you identify areas where you can improve and reduce the time and effort required for buyers to purchase and use your product. Additionally, it would be best if you strived to reduce complexity in the buying process and appoint someone responsible for conducting research and development to improve the criteria buyers use to make a purchase.

Your website is often the first point of contact between your business and potential customers, so it is critical that it is modern, easy to navigate, and features compelling messaging, calls to action, testimonials, and a clear description of what it's like to engage and buy from you.

To avoid frustrating your buyers, it is recommended that you minimize the use of chatbots and lengthy phone trees, which can come off as impersonal.

Remember that your revenue, profit, and growth come from your buyers. Therefore, it is crucial to prioritize their needs and recognize that your success depends on their satisfaction. Following these tips make it difficult for buyers to find excuses not to buy from you, and you'll be well on your way to building a thriving business.

Would you rather be Blockbuster or Netflix?

> Would you rather be a cab company or Uber?
>
> Would you rather be Kmart or Amazon?

These successful companies all followed this formula, and anyone can do it.

Over the next couple of decades, we will see a tremendous transformation in how we travel. Pay attention as to why. Electric vehicles will result in less maintenance and less dependence on oil and OPEC, and hopefully help us make our planet a cleaner place to live. Hydrogen may indeed top electricity as an auto fuel, and if it does, it will be due to being more convenient, saving time and effort finding charging stations and the time it takes to "fuel up." Drones will become more popular, perhaps for the movement of people, but very likely for transporting and delivering goods.

I have become obsessed with products and sellers and how they adopt more convenient products and buying processes. Those that save us time and effort and are more straightforward to use will and do win over those that don't. Products and sellers that figure out how to become more efficient and pass savings to buyers will beat those who settle for the status quo. Sellers' transparent pricing will beat those who hide extra charges or use fine print. Sellers who understand the importance of providing an outstanding product experience via their product or buying process will win over those who make the excuse that they cannot afford to provide a great product experience. Sellers who can convey and offer trustworthiness will continue to build a legion of devoted and faithful followers and buyers.

As a seller, it's your choice. It's not easy to provide all four, but those that get close will beat those with only one or two advantages we buyers cherish.

As we conclude, the significance of buyer-centric strategies in today's business world becomes undeniable. This chapter clearly explains how prioritizing buyer preferences in every aspect of your business can lead to sustainable growth and competitive advantage. Whether you're at the helm of a startup or an established enterprise, the lessons shared in this chapter underscore the importance of consistently aligning with buyer needs. By doing so, you position your business not just to meet but to exceed expectations, ensuring a loyal customer base and a thriving future.

Acknowledgments

Without the help of a few people, I don't think I would have had the energy to get this book across the finish line.

Michelle Gines, of Purpose Publishing set the tone and held me accountable to hitting deadlines and targets. Without her guidance, my second book would likely be just a good idea and a hope and not something that became a reality. Michelle, you are an excellent coach.

Jason Gines is a genius, and his ideation sessions with me allowed me to think through boundaries and explore how so many in this world might be able to leverage the concepts and ideas contained herein. Thank you, Jason.

My wife Pam encouraged me throughout the process and gave her precious time and talent to help me edit the final version. I love you for all you do to support me.

Rylee Gregg, who was introduced to me by a Vistage member, and a student at the University of Nebraska, did an outstanding job providing me edits and feedback for readability. Thank you for your work and dedication Rylee!

Finally, thank you to all the leaders I have engaged, especially those I have met through my CEO Peer Advisory Group. Your stories, thoughts, and wisdom enable me to see the bigger picture and appreciate your sacrifices and desires to grow so that those around you can learn and prosper.

Jim Bramlett Biography

With a career spanning over 40 years in the transportation and logistics industry, I have had the opportunity to wear many hats and contribute to the success of various businesses. A serial entrepreneur at heart, I have founded several companies, with my latest venture being a Chair at Vistage. This role allows me to guide and facilitate the growth of other business leaders, both personally and professionally.

As the founder of 5 String Solutions, a technology firm, I've dedicated our efforts to bridging the gap between shippers and local, last-mile carriers. Our unique focus on real-time information sharing and enhanced shipment visibility sets us apart in the industry.

Before delving into entrepreneurship, I held an executive role at uShip and freightPro.com and led new product development at Yellow Freight. These experiences gave me a deep understanding of the logistics and transportation sector and equipped me with the skills to start my businesses.

I have also authored two books, "The Unconventional Thinking of Dominant Companies" and "Stop The Hassle, " providing insights into how businesses can organically grow and dominate their respective markets.

My entrepreneurial journey and expertise in business growth strategy and logistics have earned me recognition in the industry. I frequently share my insights on these topics as a speaker at various events, providing valuable knowledge to the audience.

I currently reside in the Kansas City area, where I continue to explore new business avenues and contribute to the business community's growth.

When I'm not wearing my business hat, you'll find me sharing my expertise in business growth strategy and logistics or speaking at events. I call the Kansas City area home.

Made in the USA
Monee, IL
26 October 2024